International Conference on Historic Ships 2016

London, United Kingdom
7 - 8 December 2016

ISBN: 978-1-5108-8298-0

Printed from e-media with permission by:

Curran Associates, Inc.
57 Morehouse Lane
Red Hook, NY 12571

Some format issues inherent in the e-media version may also appear in this print version.

The Royal Institution of Naval Architects
8-9 Northumberland Street
London, WC2N 5DA
United Kingdom

Phone: 020 7235 4622
Fax: 020 7259 5912

publications@rina.org.uk

Additional copies of this publication are available from:

Curran Associates, Inc.
57 Morehouse Lane
Red Hook, NY 12571 USA
Phone: 845-758-0400
Fax: 845-758-2633
Email: curran@proceedings.com
Web: www.proceedings.com

CONTENTS

**Unavailable at time of going to press*

ON THE APPLICATIONS OF MODERN NAVAL ARCHITECTURE TECHNIQUES TO HISTORICAL CRAFTS

J-B R G SOUPPEZ, Southampton Solent University, UK.

SUMMARY

The Thames A Rater class had a predominant role in the popularisation of inland racing in the United Kingdom towards the end of the 19th century, and remains a historical racing class that owes its longevity to the progresses made in naval architecture and technology; the most emblematic example being the 43 feet tall carbon fibre rigs on the 27 feet centenary wooden hulls. Today, the class is a perfect illustration of the balance between historical conservation and modernisation. The design of a contemporary wooden Thames A Rater will be presented, aiming at retaining the spirit of tradition of the class, while incorporating the latest design evolutions, and complying with the current rules and regulations. Techniques such as computational fluid dynamics and parametric optimisation will be employed, leading to a significant increase in performance quantified thanks to a velocity prediction programme, thus demonstrating the applications of modern naval architecture techniques to historical crafts.

1. INTRODUCTION

To ascertain the role of modern naval architecture techniques in the conservation of traditional crafts, the design of a contemporary wooden Thames A Rater has been undertaken. This racing class appears particularly suited as its history reveals a tendency for evolution and modernisation in order to maintain the highest possible performance. The new design will make use of modern design tools, such as computational fluid dynamics, parametric optimisation and vortex lattice method, as well as consideration for the present regulations inherent to small crafts. The improvements in performance of the new A Rater will be compared to the original one thanks to the use of a velocity prediction programme, in order to demonstrate the positive impact of modern naval architecture techniques.

2. HISTORICAL BACKGROUND

2.1 THAMES SAILING CLUB

Created in 1870, the Thames Sailing Club (TSC) is the second oldest inland sailing club in Britain. The success of the first years of racing quickly highlighted two major issues: boats of highly diverse performance were competing together and no racing rules were applied. Despite those two constraints, the Thames Sailing Club became so important that in 1887, Queen Victoria herself awarded the Thames Champions Cup. This particular event revealed the potential of inland sailing events, and called for a prompt remedy to previously mentioned issues. The following year saw the creation of the Sailing Boat Association (SBA) that established racing rules, and introduced a handicap system, based on the popular Dixon Kemp's rating formula, dating 1880 [1]:

$$Rating = \frac{Lwl \times SA}{6000} \qquad Eq.\ 1$$

In which:

Lwl	Waterline length.	ft
SA	Sail area.	ft^2

This gave birth to the term '*Rater*', defining yachts designed under this particular rule; a One-Rater rating 1, a Half-Rater rating 0.5, etc. Later, a class gathering boats rating from 0.8 to 1 was created: the A Rater class.

2.2 THAMES A RATER

Towards the end of the 19th century, the design of inland racing yachts is generally defined as a '*skimming dish*', a philosophy that reached a plateau with the A Rater's fleet [2]. Out of the 13 original A Raters still racing today, twelve were built between 1898 and 1911 and the last one post WWI in 1922. The majority of the A Raters were designed by Alfred Burgoine and Linton Hope, each having a radically different approach to the rating rule that only accounts for the waterline length and the sail area.

Burgoine's yachts are characterized by a large sail area, the counterpart being a shorter waterline length. While the latter restricts the speed for a given Froude number, the larger sail area will offer a more powerful boat that therefore has to be made wider to increase form stability and the ability to carry sail. On the other hand, Hope favoured a longer waterline and narrower beam, and consequently a smaller sail area as dictated by the rating rule. The opposition of those two design philosophies is illustrated in Table 1, comparing two original A Raters, namely *Ulva* (1898) and *Scamp* (1902), respectively designed by Burgoine and Hope.

Table 1: Burgoine and Hope designs comparison

Yacht	Lwl (m)	Bwl (m)	SA (m²)	Rating
Ulva	4.80	2.15	35.00	0.99
Scamp	5.15	1.66	33.00	1.00

The radically opposed specifications led to distinctive performances, the Hope's yachts being better suited to upwind sailing while the Burgoine's ones sailed faster downwind.

2.3 CHEATING THE RULE

Looking at the various attempts to *cheat* the A Rater class rules provides some insights into the critical design areas to be improved; in this case the waterline length, the stability and the mast weight.

Firstly, the work of William Froude published a few decades before the A Raters [3, 4] identified the waterline length as the main speed restricting factor, hence the interest in a longer waterline length. For the typical resistance hump occurring at a Froude number of 0.33, Ulva would achieve 4.40 knots, while Scamp would reach 4.56 knots. As a result, some boats were fitted with rods and wires at each end. By winding up the wires, the yacht could artificially be sagged to offer a shorter waterline length when measured. The wires would then be loosened when racing, thus extending the actual waterline length.

Secondly, stability is a major factor for such a light displacement craft carrying a large sail area. Some of the main innovations with regard to stability have been experimented on Vagabond, designed in 1907 by Hope. The ancestor of the trapeze was named the *'bell rope'*: a crew, the *'bell boy'*, holding onto a rope attached at the top of the mast could stand to windward, as depicted in Figure 1, thus increasing the righting moment.

Figure 1: The 'bell boy' and the 'bell rope' [2]

Once made illegal, Vagabond was fitted with sliding seats (see Figure 2), with the same effect of increasing the righting moment, and the same fate of being banned.

Figure 2: Sliding seats on Vagabond [2].

Finally, removable top masts were introduced to minimise the heeling moment in high winds. While this practice was prohibited, the masts would undergo several improvements in the future.

2.4 EVOLUTION

While the hulls and appendages have been untouched since the beginning of the 20th century, the rig and sails have significantly evolved. Originally designed as a low aspect ratio gaff rigs, the masts evolved from bamboo to the current carbon fibre, via solid and hollow wooden spars as well as aluminium. With the improving mast technology, higher spans could be achieved, and the A Raters are now famous for their impressive 43 feet (13.1m) tall rigs, depicted in Figure 3.

Figure 3: Rig in 1907 (left) [2] and 2014 (right) [5].

One of the downsides of the early gaff rigs was the eccentric location of the centre of effort of the sails downwind, requiring tremendous efforts from the helmsman to keep the boat on course in the narrow waterways. Remains of this behaviour can be seen today with some of the original tillers, clearly made for the helmsman to hold onto it firmly, as shown in Figure 4.

Figure 4: The bracing tiller of Ulva [6].

Along with the masts, cotton sails have been replaced with more advanced materials. Those innovations contributed to the success of the A Rater class, and so did the Glass Reinforced Plastic (GRP) technology that sparked a regain of interest in the class in the late 1970s.

2.5 MODERN DAYS

The A Rater class is one of the rare racing classes that survived after World War II, but with the last wooden A Rater dating from 1922, the number of boats was becoming smaller and smaller over time. In 1978, a female mould tool of *Ulva* was made, and new GRP hulls were built, thus ensuring the future of the class. Around this time also came into place a change in the rules: no new design would be allowed, any new A Rater would have to be an exact replica of an original one; as stated by the Thames A Rater class rule [7] and further discussed in Section 3.1. In addition, to stop the *arms race* resulting from the new composite manufacturing, a minimum class weight was imposed.

The early 2010s saw the appearance of the first full carbon boats, fitted with a new deck inspired from the 5o5 class, and thus moving away from the traditional designs; the latest A Rater built is pictured in Figure 5.

Figure 5: The latest A Rater built [5].

The challenge set is to design the next generation of wooden A Raters, marrying performance with tradition.

2.6 DESIGN BRIEF

The design brief aimed to involve as many stakeholders as possible, thus considering the requirements of the client, shipyard, and sailor's feedback. All agreed to an aesthetically pleasing wooden yacht, with classic lines to carry the historical legacy of the A Rater class, with however a stronger emphasis on improved performance.

Additional considerations originated from the environmental constraints. Due to the specific area of operation of the craft, namely the Norfolk Broads, as opposed to the Thames, requirements for the maximum draft and bridge heights had to be taken into account [8]. Furthermore, the Thames A Rater class rule requirements [7], detailed in Section 3.1, were incorporated.

Finally, in terms of the regulatory framework, the boat being aimed at the European market, it is to comply with the Recreational Craft Directive (RCD). As a consequence of the move to the RCD II [9] in January 2017, the yacht was designed to the newer regulation, and in accordance with the relevant ISO standards.

2.7 CONCLUSIONS

The historical Thames A Rater class is a significant part of England's inland racing history, and is still a vibrant class nowadays, primarily thanks to the compromise between tradition and evolution. The acceptance of the design and technology progresses made in rig and sails allowed the A Rater class to carry its legacy through time. And this is the objective for the new design: a high performance craft building on the latest design and technology, while preserving the spirit of tradition of the class; thus utilising modern naval architecture techniques to support the conservation of a historical class.

3. HULL DESIGN

3.1 CLASS RULE

The Thames A Rater class rules [7] specifies the requirements for a craft to meet the one design rule. This ranges from a minimum lightship weight of 750 lbs (340kg), to a maximum mast height from the sheerline of 43ft, and a sail area of 350 ft^2 (32.51m^2). But the primary design constraint is given by rule D2 [7]:

"D2 New Yachts

A new hull will only be considered to be an A class rater hull if it is an exact replica of an existing Rater as defined above, taken from either an existing hull, or original lines, subject in both cases to a tolerance of one and one half inches."

While some linesplans are still in existence, owners are very protective of those. The linesplan of an original Thames A Rater is therefore to be found in the public domain in order to provide the basis of the new yacht. The only publicly available linesplan is featured in the 11th edition of *Dixon's Kemp manual of yacht and boat sailing* [60], reviewed by Linton Hope, who added the linesplan of the Thames A Rater *Scamp*.

3.2 MODELLING SCAMP

Designed in 1902, *Scamp* has always been a successful boat, and being one of the original Thames A Rater, it qualifies as an exact replica of an existing A Rater and will therefore be adopted as the basis hull of the new design.

3.2 (a) Taking the Lines

When dealing with one of the last drawings of an historic craft, the priority is to ensure the integrity of the document and avoid any form of damage to it. With this is mind, the state-of-the-art facilities available at the British Library have been utilised to obtain a digital copy of the linesplan, as shown in Figure 6.

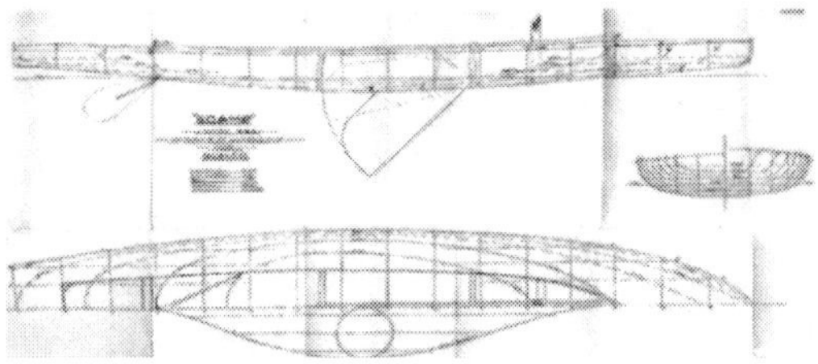

Figure 6: Original linesplan of Scamp (1902) [10].

Unfortunately, the drawing was slightly distorted due to the folds and the deformation due to aging. While it constitutes a good graphical representation, it does not allow for an accurate enough modelling of the boat.

As a result, the lines were manually taken off by physical measurements of all the offsets to the closest 1/64th of an inch (accuracy of ± 1/128th of an inch). The lines were taken solely from the body plan. Indeed, since the body plan was drawn over a small area, it had been less affected by distortion and aging or folds compared to the half-breadth and profile view extending the full length of the plan.

3.2 (b) 2D Drawing

The table of offsets realised was then scaled up to full size, converted from imperial to metric, and numerically lofted using computer aided design (CAD). This process enabled to redraw the 2D linesplan, ensuring an exact replica is achieved, as shown in Figure 7.

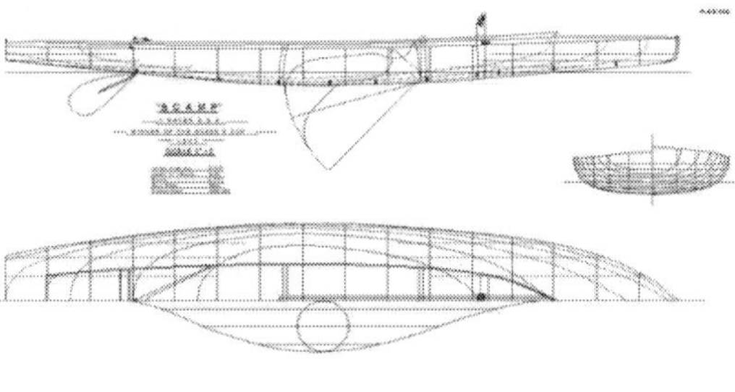

Figure 7: Replica of the Scamp linesplan

Note that this linesplan is an exact replica of the original one, reproducing every detail, even where discrepancies have been noticed, as it is the case with the centreboard, later discussed in Section 4.2 (a).

3.2 (c) 3D Modelling

Scamp was then modelled in 3 dimensions, in a process very similar to the one of traditional boatbuilding. First, the stations are positioned along the length of the craft; a surface is then lofted along those stations with a specified accuracy of 0.01 mm. The process can be observed in Figure 8.

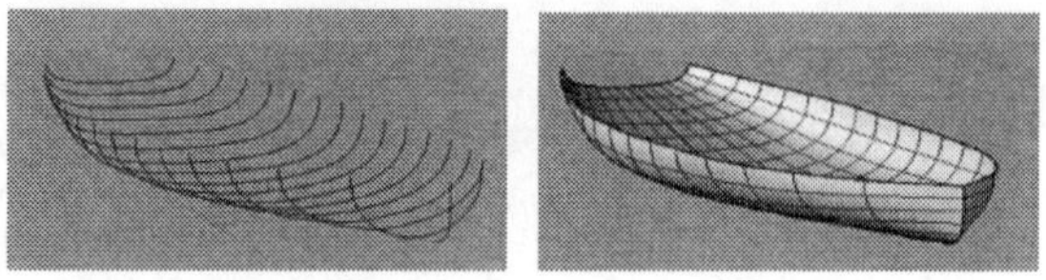

Figure 8: 3D modelling of Scamp.

The hull surface then allows to ascertain the hydrostatics.

3.2 (d) Hydrostatics

The hydrostatics of the 3D model have been compared to those determined from the replica of the 2D linesplan using Simpson's rule. The results in Table 2 reveal a very accurate modelling, with an average 0.46% difference, well within the uncertainty inherent to each method.

Table 2: Hydrostatics comparison

Parameter	Linesplan	3D Model	Diff.	Diff. (%)
LOA (m)	8.28	8.28	0.000	0.00%
Lwl (m)	5.15	5.17	0.019	0.37%
BOA (m)	1.90	1.90	0.000	0.00%
Bwl (m)	1.66	1.64	-0.020	-1.20%
Tc (m)	0.16	0.16	-0.002	-1.25%
F (m)	0.31	0.31	0.000	0.00%
Disp. (m^3)	0.548	0.545	-0.003	0.00%
Awp (m^2)	6.58	6.60	0.020	0.30%
LCB (m)	2.84	2.80	-0.043	-1.53%
LCF (m)	2.81	2.78	-0.033	-1.16%
Cb	0.40	0.40	-0.003	-0.71%
Cp	0.59	0.59	-0.006	-1.04%
Cm	0.68	0.68	0.002	0.34%

An exact replica of *Scamp* has therefore been achieved, thus complying with the class rule, which does however allow a one and a half inches (31.8mm) building tolerance [7]. Due to the reliability of modern wood construction, and the new boat being cold-moulded, part of that tolerance will be utilised to modify and improve the hull design while still meeting the rule.

3.3 DESIGN MODIFICATIONS

3.3 (a) Design Tolerances

In order to establish how much of the tolerance can be allocated to modify the design, the accuracy lost in modelling the hull from the original linesplan must be evaluated, and building tolerances must be estimated.

By taking the lines off, the accuracy was maximised compared to simply drawing over the top of the linesplan that has been distorted over time. As a result, the replica of the linesplan can be considered as accurate as the measurement tolerance, i.e. 1/128th of an inch (0.20mm), which translates into 1/128th of a foot (2.38mm) full size.

To this must be added the 3D modelling accuracy of 0.01mm, giving a total uncertainty of 2.39 mm.

While a high standard of manufacturing can be expected, it is important to allow for slight building inaccuracies, such as the lofting of the station moulds, the natural expansions and contractions of the wood, and the possibility of a human error. An overall margin of 3/8th of an inch (9.53mm) has therefore been allocated to the construction, leaving 19.88mm to modify the hull shape. This value has been rounded up to 20mm, thus decreasing the building tolerance to 9.41mm.

The 20mm margin offers an opportunity to improve the hull design. It is to be noted that, while the overall one and a half inches tolerance will be respected, it is in practice quite hard to enforce this aspect of the class rule. Indeed, while some physical dimensions such as the length and breadth of the boat can physically be measured, a change in curvature in the middle of a section cannot however be clearly identified. This is supported by the International Towing Tank Conference (ITTC) standard for towing tank models manufacturing [11], where a tolerance is allowed for the length, breadth and depth of the model, while other deviations of the hull shape from the intended geometry are neglected due to the impracticalities of comparing the two.

3.3 (b) Empirical Resistance Model

The 20mm allocated to design modification will aim at improving the hull shape by reducing its resistance thanks to parametric optimisation based on the Delft Systematic Yacht Hull Series (DSYHS) [12].

The DSYHS offers regression equations that enable to assess the resistance of a yacht from its principal dimensions. The two main drag components of a bare hull are namely the frictional and the residuary resistance.

At low Froude numbers (slow speeds), the frictional resistance is the major component, and is dependent on the wetted surface area of the hull (and inherent roughness). At higher Froude numbers, the residuary resistance becomes the primary drag component. In this instance, efforts have been focussed on decreasing the resistance at high Froude numbers due to the high speeds and planing ability of the A Raters.

The upright hull residuary resistance Rrh is given by [12]:

$$\frac{Rrh}{\nabla c \times \rho \times g} = a_0 + \left(a_1 \times \frac{LCB_{fpp}}{Lwl} + a_2 \times Cp + a_3 \times \frac{\nabla c^{2/3}}{Aw} + a_4 \times \frac{Bwl}{Lwl} + a_5 \times \frac{LCB_{fpp}}{LCF_{fpp}} + a_6 \times \frac{Bwl}{Tc} + a_7 \times Cm \right) \times \frac{\nabla c^{1/3}}{Lwl} \quad \text{Eq. 2}$$

In which:

Rrh	Residuary resistance.	N
∇c	Canoe body displacement.	m^3
ρ	Water density.	$kg.m^{-3}$
g	Acceleration due to gravity.	$m.s^{-2}$
LCB_{fpp}	LCB location from the FPP.	m
Lwl	Waterline length.	m
Cp	Prismatic coefficient.	-
Aw	Waterplane area.	m^2
Bwl	Waterline beam.	m
LCF_{fpp}	LCF location from the FPP.	m
Tc	Canoe body draft.	m
Cm	Midship area coefficient.	-
a_0 to a_7	Regression coefficients.	-

3.3 (c) Parametric Optimisation

Thanks to the DSYHS, the most influent parameters on the resistance can be identified. However, the regression coefficients (a_0 to a_7 in Eq. 2) have a varying influence between displacement mode and semi-displacement mode; the sailing regime transition occurring around a Froude number of 0.45 [13].

The hull of the original *Scamp* has therefore been modified based on parametric optimisation to reduce the resistance at higher Froude numbers. The objective was reached with an average 3% reduction in overall resistance in a fully loaded condition (replicating the sailing displacement) past a Froude number of 0.45, as illustrated in Figure 9.

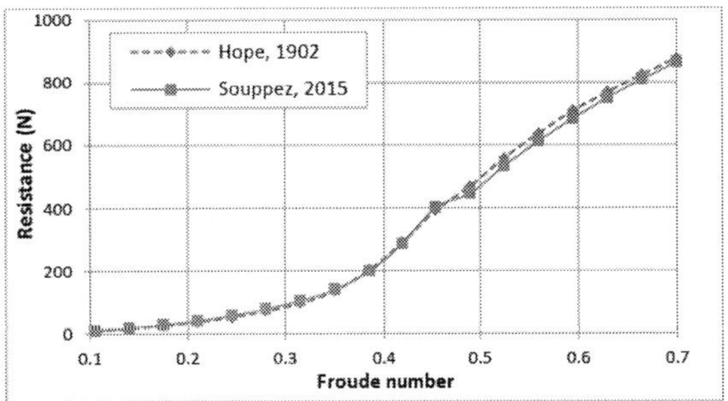

Figure 9: Resistance comparison.

The reduction is visible with the sudden discontinuity in the resistance curve occurring at a Froude number of 0.45. A similar resistance decrease was also observed at all sailing heel angles.

3.3 (d) Modified Scamp

The main modifications realised using the 20mm margin include an extended overall length and breadth, to respectively increase the Froude number and form stability. The angle of the bottom of the boat with the waterline being so acute, an additional 20mm of overall length resulted in an impressive 160mm increase in waterline length.

The draft has been shortened to provide a shallower and flatter craft; effectively increasing the values of the midship and prismatic coefficients, desirable for the purpose of the parametric optimisation and planing capabilities.

The modifications induced a 6.19% increase in wetted surface area, which implies a slight increase in frictional drag, largely over-compensated by the resistance reduction at higher Froude number. The comparison between the original Hope design and the modified *Scamp* is summarized in Table 3.

Table 3: Original and new Scamp design comparison

Parameter	Hope	Souppez	Diff. (%)
Length over all (m)	8.28	8.30	0.24%
Length on waterline (m)	5.46	5.62	2.93%
Beam over all (m)	1.90	1.92	1.05%
Beam on waterline (m)	1.67	1.70	1.80%
Canoe body draft (m)	0.185	0.166	-10.27%
Displacement (m^3)	0.652	0.652	0.00%
Midship coefficient	0.734	0.762	3.81%
Prismatic coefficient	0.528	0.546	3.41%
Wetted surface area (m^2)	7.112	7.552	6.19%
Waterplane area (m^2)	6.887	7.330	6.43%

3.4 CONCLUSIONS

The original linesplan of an existing Thames A Rater has been considered for the new design, thus complying with the class rule. *Scamp* was accurately redrawn in 2D and then converted to 3D. By accounting for the modelling uncertainties and estimating the building tolerance required, a 20mm margin has been employed to improve the hull shape. Thanks to parametric optimisation, the resistance at high Froude numbers was reduced by 3%, and enhancements in terms of hydrostatics and stability have been achieved, giving the final hull design for the new A Rater.

No further work has been conducted on refining the hull with other methods, such as computational fluid dynamics (CFD). Indeed, a large amount of time would be required to only provide very minor improvements. Instead, the work has been focussed on the appendages design: since the centreboard and rudder do not have to be replicas, they offer a great opportunity to elaborate the hydrodynamics and will concentrate the majority of the development efforts.

4. APPENDAGES

4.1 COMPUTATIONAL FLUID DYNAMICS

4.1 (a) Modelling and Simplifications

CFD has been employed to provide an initial comparison of a range of planforms for both the centreboard and rudder. The modelling has been heavily simplified: the presence of the hull and free surface and inherent impact on the appendages have been neglected. The appendage tested is therefore modelled alone in a domain of water; this is motivated by the restricted computational power available. All simulations have been performed with a Reynolds-Averaged Navier-Stokes Equations (RANSE) solver.

Since the structural and manufacturing constraints are not known at this stage, the appendages have been treated as thin flat plates. Indeed, the actual section will have to consider the structural requirements, thus dictating the thickness/chord ratio.

4.1 (b) Governing equations

The steady state analysis performed is to satisfy both the continuity (conservation of mass) and the momentum (conservation of linear momentum) equations, respectively given by:

$$\nabla.(\rho\langle u_k\rangle) = 0 \qquad \text{Eq. 3}$$

And:

$$\nabla.(\rho\langle u_k\rangle\langle u_k\rangle) = -\nabla\langle p\rangle + \nabla.\overline{\tau_k} + \rho\vec{g} \qquad \text{Eq. 4}$$

Where:

$$\overline{\tau_k} = \mu_{eff}\left(\nabla\langle u_k\rangle + (\nabla\langle u_k\rangle)^T - \frac{2}{3}\mu_{eff}\nabla.\langle u_k\rangle I\right) \qquad \text{Eq. 5}$$

The k-ε turbulence model [14] has been chosen as it is an industry standard and has been extensively studied [15]. For the purpose of this analysis, a typical turbulence intensity I_T of 5% has been considered. The model is governed by two equations, the turbulence kinetic energy and energy dissipation rate:

$$\nabla.((\langle u_k\rangle\rho k) = \nabla.\left[\left(\mu + \frac{\mu_t}{\sigma_k}\right)\nabla.k\right] + G_k \qquad \text{Eq.6}$$

And:

$$\frac{\partial(\rho\varepsilon\langle u_k\rangle)}{\partial x_k} = \frac{\partial}{\partial x_k}\left(\left(\mu + \frac{\mu_t}{\sigma_k}\right)\frac{\partial\varepsilon}{x_k}\right) + C_{\varepsilon 1}\frac{\varepsilon}{k}G_k - C_{\varepsilon 2}\frac{\varepsilon^2}{k} \qquad Eq.\ 7$$

Where:

$$G_k = \tau_{ik}\frac{\partial\langle u_i\rangle}{\partial x_k} \qquad \sigma_k = 1.00$$

$$\sigma_\varepsilon = 1.3$$

$$\nu_T = C_\mu\frac{k^2}{\varepsilon} \qquad C_{\varepsilon 1} = 1.44$$

$$C_{\varepsilon 2} = 1.92$$

$$C_\mu = 0.09$$

4.1 (c) Domain Size

The domain size, normally expressed in terms of boat length L has been kept relatively small to minimise computation time. The dimensions are: 2L long (0.5L upstream, 1.5L downstream), 0.5L wide, and 0.5L deep. Investigations into wider and deeper domains proved not to impact the results, suggesting that the selected domain is wide and deep enough to avoid blockage issues.

4.1 (d) Mesh

An unstructured mesh was adopted as it is easier and faster to create [15]. In addition, inflation layers have been built around the surface of the appendage to better capture the boundary layer and inherent viscous components.

To maximise accuracy while minimizing computational resources, a mesh convergence study has been conducted; results are shown in Figure 10.

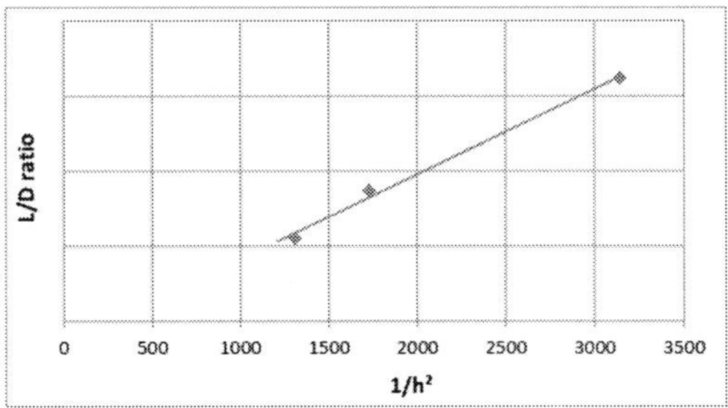

Figure 10: Convergence study.

The results reveal the second order nature of the solver (straight line achieved for $1/h^2$), and suggest that the region of monolithic convergence has been reached.

4.1 (e) Error Estimation

The discretization error and grid convergence index (GCI) have been evaluated following the Richardson extrapolation procedure and the Roach and Celik error estimation [16]. The results are summarized in Table 4.

Table 4: Discretization error

Discretization Error Results						
r_{21}	r_{32}	p	F^{12}_{ext}	e^{21}_{a}	e^{21}_{ext}	GCI^{21}
1.35	1.35	2.58	10.25	0.29%	0.25%	0.32%

The negligible errors demonstrate the level of accuracy achieved with the finer mesh; no further refinement appears to be necessary. Indeed, a finer mesh would not significantly increase the accuracy, but induce a tremendous increment in solving time.

4.1 (f) Boundary Conditions

The boundary conditions set are described in Table 5.

Table 5: Boundary conditions

Boundary	Applied Condition
Upstream end	Inlet, specified U velocity.
Downstream end	Outlet, specified U velocity.
Surrounding walls	No slip, smooth walls, U velocity.
Centreboard	No slip, smooth wall, stationary.

4.1 (g) Applications

In order to provide an initial comparison between a range of planforms using a RANSE solver, the problem had to be simplified due to the limited computational resources available. The method presented in this section will be applied to the centreboard and rudder design, respectively tackled in Section 4.2 and Section 4.3.

4.2 CENTREBOARD

4.2 (a) Design Discrepancy

The linesplan of *Scamp* offers two illustrations of the centreboard: one lowered under the hull (*centreboard 1*), and one retracted inside (*centreboard 2*). Careful analysis of the drawing revealed that those two centreboards are not identical, as depicted in Figure 11.

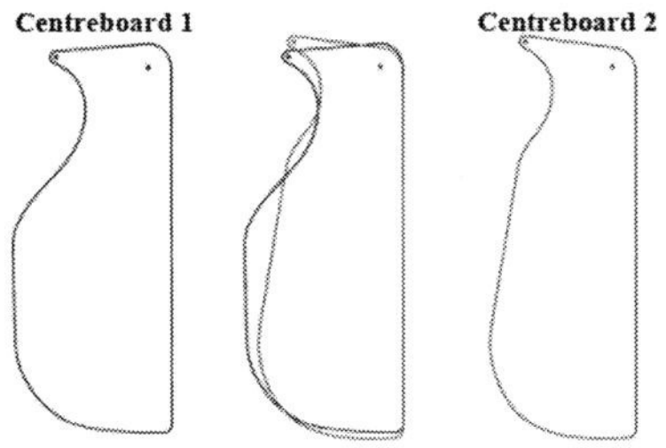

Figure 11: Centreboards discrepancy.

This difference is too extreme to simply result from a deformation of the plan, and is assumed to be a drawing error. Whilst it is impossible to know with certainty which one is the intended planform and which one is the mistake, two arguments would suggest that *centreboard 1* is the original design.

Firstly, *centreboard 1* is drawn as a solid thick line in the lowered position, where the centreboard is to operate. Conversely, *centreboard 2* is a thin dotted line that would typically be drawn as a rotation of the lower one, and therefore prone to a drawing mistake.

In addition, the choice of the appendages area was often taken as a percentage of the sail area; an approach still employed by many designers today.

Centreboard 1 represents 2.30% of the sail area, which when added to the rudder area gives a total 3.00%, both round numbers suggesting they are the intended proportions. Conversely, *centreboard 2* has an area equal to 2.12% of the sail area, as summarised in Table 6.

Table 6: Appendage areas.

Appendage	Area (m²)	% of Sail Area
Centreboard 1	0.76	2.30%
Centreboard 2	0.70	2.12%
Rudder	0.23	0.70%

There is therefore evidence to suggest *centreboard 1* is the intended planform, but no certainty, hence both centreboard designs will be considered and analysed.

4.2 (b) Area

The planform area can lead to a large loss of performance if too little or too much is provided, respectively resulting in high leeway angle and added frictional resistance. In this case, one option could be to keep the original design area (2.30%). Alternatively, an empirical estimate can be calculated [17]:

$$\frac{KA}{SA} = 0.039 \times \frac{Tk}{LOA} + C \qquad Eq.\ 8$$

In which:

KA/SA	Keel/sail area ratio.	%
Tk	Keel draft.	m
LOA	Length overall.	m
C	0.018 for racing yachts.	-

In this case, a 2.51% ratio is obtained.

Finally, a more advanced method [18] based on the area required when coming out of a tack in light winds has been applied. A keel area of 2.27% of the sail area has been ascertained, i.e. very close to the original 2.30%. The later value has therefore been conserved.

4.2 (c) Comparative Study

Having established the keel area (2.30% or 0.76m²), and based on the draft restriction of the area of operation [8], a range of possible designs have been investigated, based on the setup outlined in Section 4.1. Indeed, the class rule does not specify that the appendages have to be replicas of the original, thus allowing for hydrodynamic improvements. The centreboard designs considered are briefly described in Table 7 and illustrated in Figure 12.

Table 7: Centreboard designs.

Design	Characteristics
Scamp 1	Centreboard 1 [10].
Scamp 2	Centreboard 2 (see 4.2 (a)) [10].
Sorceress	L. Hope design for *Sorceress* [10].
Semi-Elliptic	Elliptical loading (cf. Spitfire wings).
Rectangular	Easiest to manufacture.
Inverted	Increased ballast lever (cf. *Australia II*).
Stewart	Stewart modern A Rater design [19].
Scow	Found on modern scows (cf. *E-Scow*).
45% Taper	Hydrodynamic optimum taper ratio [20]
20° Swept	Popular dinghy design [20].

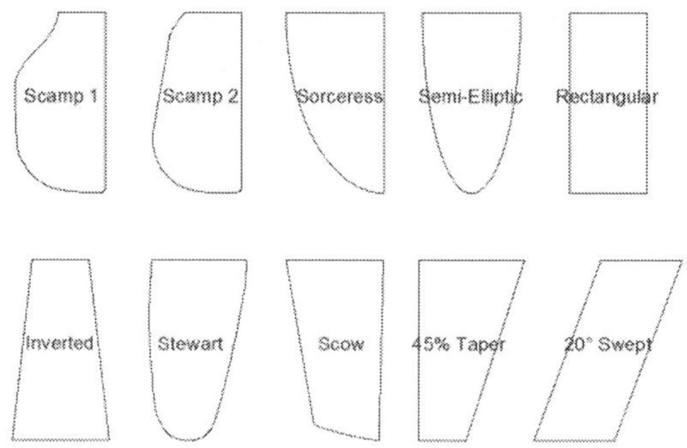

Figure 12: Centreboard designs investigated.

The analysis has been performed at 15° of heel and 5° of leeway, for speeds of 1, 2, 3, 4 and 5 m/s, representative of typical sailing conditions.

4.2 (d) Results

The deltas (change) in lift/drag ratio compared to the original centreboard (Scamp 1), are shown in Figure 13.

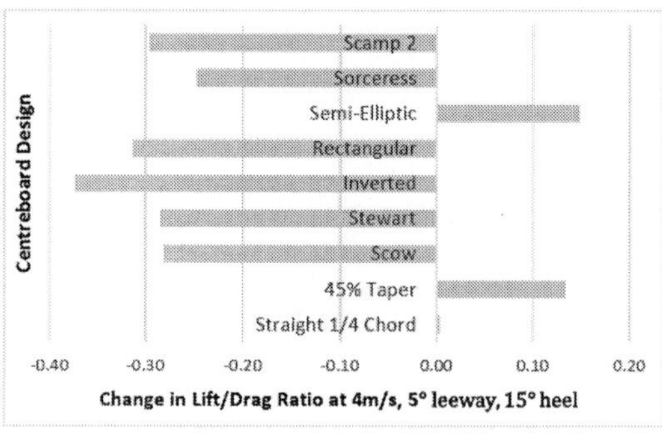

Figure 13: Centreboards delta in Lift/Drag ratio.

The overall design ranking is as expected: the semi-elliptic and 45% taper ratio, both promoting elliptical spanwise loading, proved to be the most efficients, especially at low speeds, due to a large reduction in tip vortex. The better hydrodynamic performance of the semi-elliptic centreboard constitutes a 1.8% improvement in terms of lift/drag ratio compared to the original.

4.2 (e) Section

The foil section of yachts are generally NACA 00 series, due to their better hydrodynamic performance, with a higher lift/drag ratio and a delayed stall angle compared to NACA 63, 64, and 65 series. A closer analysis revealed further advantages of the NACA 00 series.

Indeed, as detailed in Table 8, a NACA 00 series has the highest sectional area coefficient. Consequently, the ballast/WSA ratio will be the highest, meaning a lower wetted area and frictional resistance for a given volume.

Moreover, the section modulus/WSA ratio is also the highest, hence for a given structural requirement, a NACA 00 will again have a lower wetted area and thus less drag.

Table 8: NACA series comparison

Section	Area Coeff.	Ballast/WSA	SM_T/WSA
NACA 0012	0.676	100%	100%
NACA 63 012	0.659	97.65%	95.19%
NACA 64 012	0.628	92.90%	86.30%
NACA 65 012	0.644	95.29%	90.79%

The thickness/chord ratio was dictated by the structural constraints from the ISO 12215-9 [26] and varies from the root to the tip due to the decreasing bending moment, with an average value of 7.2%.

4.2 (f) Conclusions

Based on the hydrodynamic analysis, the semi-elliptic centreboard appears to be the most efficient, and will therefore be the proposed design. In addition, due to its geometrical properties, the centreboard offers one of the highest centres of lateral resistance, thus decreasing the heeling arm, contributing to an improved stability. A NACA 00 section has been chosen and an average 7.2% thickness/chord ratio was defined based on the structural analysis.

4.3 RUDDER

4.3 (a) Area

As per the centreboard, the rudder area is generally expressed as a ratio of the sail area. For small yachts, 1.5% is usually advised [20]. This value however appears to be excessive for the A Rater *Scamp*, having less than half the area suggested.

Comparing the appendages of *Sorceress* (1894) and *Scamp* (1902), both designed by Linton Hope, showed that both the rudder and centreboard areas had been decreased, suggesting that only a minimum area is required. As a result, the original rudder area of 0.23m², or 0.70% of the sail area, will be kept. The rudder planform will however be redesigned.

4.3 (b) Comparative Study

Using the same CFD approach as for the centreboard, various rudder planforms will be investigated; aiming at increasing the rudder aspect ratio to improve the hydrodynamic performance. The ten designs testd are depicted in Figure 14 and characterised in Table 9.

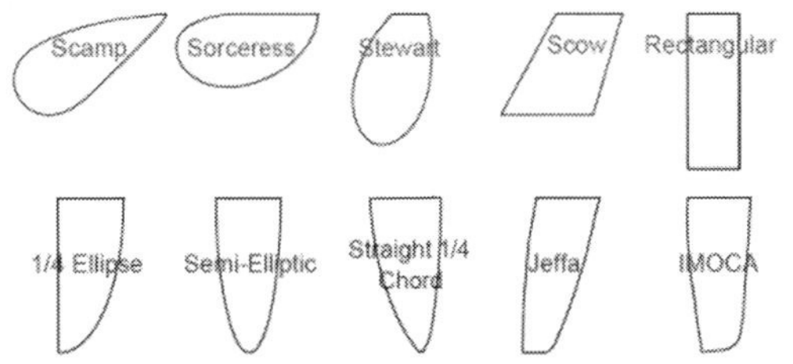

Figure 14: Rudder designs investigated.

Table 9: Rudder designs.

Design	Characteristics
Scamp	Original *Scamp* rudder [10]
Sorceress	L. Hope design for *Sorceress* [10].
Stewart	Stewart modern A Rater design [19].
Scow	Found on modern scows (cf. *E-Scow*).
Rectangular	Easiest to manufacture
1/4 Ellipse	Popular hydrodynamic design [20].
Semi-Elliptic	Elliptical spanwise loading.
Straight 1/4 chord	Popular hydrodynamic design [20].
Jeffa	Typical modern cruiser/racer design.
IMOCA	Typical modern fast racing design.

4.3 (c) Results

The analysis has been performed at 15° of heel and 5° of leeway, for speeds of 1, 2, 3, 4 and 5 m/s, as per the centreboard, with however different results due to the change in aspect ratio. The differences in lift/drag ratio compared to the original rudder (Scamp) are shown in Figure 15.

The straight ¼ chord rudder appears to provide more lift and less drag than the other planforms tested, and led to a 10.5% improvement compared to the original. Those results are supported by tests realised for a rudder angle of 15° (close to stall), where the straight ¼ chord also proved to be the most efficient, as it provides a significant decrease in tip vortex compared to the original *Scamp* rudder.

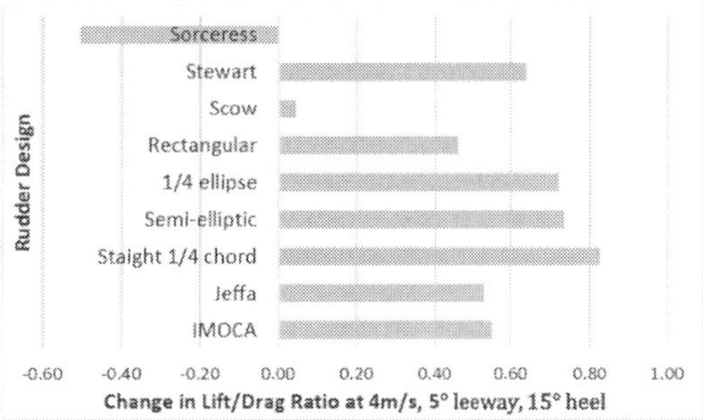

Figure 15: Rudders delta in Lift/Drag ratio.

4.3 (d) Section

A NACA 00 series has been selected due its higher stall angle compared to other NACA series. A thin foil would have minimum drag but a low stall angle. Conversely, a thick foil would delay stall, but increase the drag.
A 12% thickness/chord ratio is often seen as a good compromise, and was adopted in this case. Moreover, the chosen section also proved to be structurally suitable and able to accommodate the rudder stock diameter required to satisfy the ISO 12215-8 [25].

4.3 (e) Conclusions

Following a comparative CFD analysis, a straight ¼ chord line rudder configuration has been selected, with a standard NACA 0012 foil section.

4.4 CONCLUSIONS

A basic CFD setup has be used to contrast a range of planforms for both the centreboard and rudder, aiming at a maximum lift developed for a minimum drag. A semi-elliptic planform was therefore chosen for the centreboard, while a straight ¼ chord proved to be the most efficient rudder. The new centreboard and rudder improved the lift/drag ratio by 1.8% and 10.5% respectively compared to the original design. However, further considerations such as the location of the appendages will be required to achieve a better balance between the aerodynamic centre of effort of the sails and the hydrodynamic centre of lateral resistance of the appendages, detailed in Section 6.2 (b).

5. COCKPIT DESIGN

5.1 INTRODUCTION

The design of a practical, polyvalent, and safe cockpit is a major requirement for open boats. The three main cockpit design philosophies are depicted in Figure 16, and are namely:

- A narrow rectangular cockpit, such as the original *Scamp* (left).
- A slightly wider cockpit running parallel to the sheer (centre).
- A very wide cockpit (right).

Figure 16: Cockpit design philosophies [7].

The pros and cons will be highlighted in order to find a compromise, with considerations for downflooding, ergonomics and anthropometrics.

5.2 SIDE DECK WIDTH

5.2 (a) Downflooding Angle

The foremost consideration inherent to the downflooding angle is the width of the side decks, and resulting width of the cockpit. On original A Raters such as *Scamp*, the governing factor was avoiding water intake when heeling over or capsizing. The angle at which water gets inside the cockpit is known as the downflooding angle. As presented in Figure 17, the downflooding angle of *Scamp* is just above 90°, thus coherent with a safe design to avoid water intake, and potentially sinking the boat when capsizing.

Nowadays, the buoyancy aid imposed by the A Rater class rule [7] prevents sinkage, leading to much more open cockpits. This allows more space for manoeuvers, and the crew can provide an increased righting moment without having to hike yet since they are sitting further out. The drawback is an earlier water intake, potentially within operating heeling angles, which would handicap the performance of the boat.

The proposed design offers a good compromise, with a 350mm wide side deck, which allows both a comfortable and efficient position of the crew, while retaining a high downflooding angle of 48°, thus satisfying the relevant ISO regulation [21], as demonstrated in Figure 17.

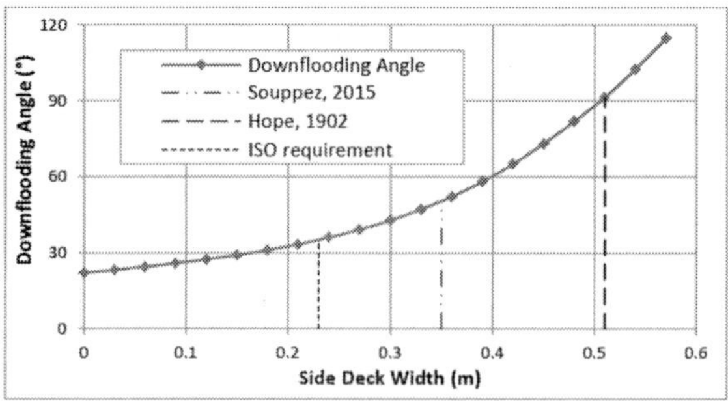

Figure 17: Downflooding angle.

5.2 (b) Ergonomics and Anthropometrics

Another decisive factor, unfortunately rarely incorporated in deciding upon the width of the side deck is the location of the sailor's back-knee. Consequently, both ergonomics and anthropometrics have been taken into account, respectively defined as the science of designing spaces and environments and the study and measurements of human proportions.

From a comfort point on view, any edge resting on the back-knee when hiking is particularly painful, and the position cannot be sustained for prolonged amount of time. Furthermore, experiments realized on improving hiking positions [22] revealed that the shorter the back-knee/sheer distance, the more efficient and comfortable the hiking is.

With the proposed cockpit, the sheer lies just above the back-knee when fully hiked, for a prolonged and efficient contribution to the righting moment. Plus, the inner edge of the cockpit will also lie just above the back-knee of a crew seeking support on the centreboard case.

The width of the cockpit and resulting side decks width have therefore been fixed based on both downflooding angle and comfortable hiking. The length of the deck however is governed by a different factor.

5.3 DECK LENGTH

Longitudinal strength is a likely issue for a yacht such as the A Rater, as confirmed by the ISO 12215-6 guidance [23]. Thus the deck offers a significant structural contribution not to be neglected. Indeed, considering the entire boat as a beam, the deck is located away from the neutral axis, hence its predominant contribution.

On the one hand, a small open space has been retained in front of the mast in order to provide a convenient location to store excess ropes or gear, without it overcrowding the cockpit.

On the other hand, the aft deck has been extended forward of the rudder stock; this will also provide a stronger support point for the top rudder bearing, thus improving the steering comfort.

However, the deck cannot be fully closed, as it must provide sufficient space for the crew to perform manoeuvres. Furthermore, the mass of the crew constitutes a large proportion of the overall weight; a large cockpit offering a wide range of longitudinal positions for the crew will therefore improve the sailing equilibrium. To encourage this, the compartments of the cockpit have been redefined.

5.4 COMPARTMENTS

The cockpit of the original *Scamp* is divided into three main areas, one for each crew. The helmsman has the largest, while the forward crew is confined into a very small space. This is in contradiction with a typical race crew organization, where the helmsman would remain mostly static, while the two other crew would be in charge of the balance. It is therefore crucial to have the ability to move forward and aft so that the longitudinal balance of the yacht can be adjusted, particularly to promote planing.

The new cockpit is only composed of two compartments: the aft one dedicated to the helmsman, and a spacious forward one for the two crew members. As a result, the sailors can move further forward and further aft, while still having all control lines in close proximity.

The definition of the compartments is primarily dictated by the location of the two main structural bulkheads. The first one, in way of the mast, has to withstand the mast, shrouds, and centreboard loads. The second one, separating the two main cockpit areas, supports the aft end of the centerplate case, the mainsheet, traveller and running backstays.

5.5 CONCLUSIONS

Starting from the original cockpit, an improved version has been designed, with a smaller side deck width to provide a more comfortable hiking position, while retaining a high downflooding angle. In addition, the overall length of the cockpit has been decreased for structural purposes, the actual usable volume has however been greatly increased. Finally, a new layout has been created, offering a larger compartment for the two forward crew members. A graphical comparison of the original and new cockpit designs is presented in Figure 18.

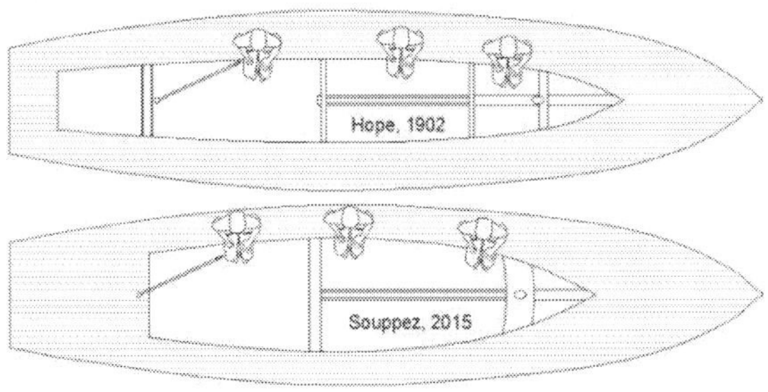

Figure 18: Cockpit layout comparison.

6. DESIGN EVALUATION

6.1 COMPLIANCE

The vessel was designed for the RCD II, taking into account the relevant harmonised ISO standards. All requirements have been satisfied for category C, inshore, rather than category D, inland. This is primarily motivated by the extreme nature of the Thames A Rater, thus providing an additional factor of safety; but also to extend the vessel's programme to inshore sailing, thus widening the area of operation.

6.1 (a) Structural Design

In order to fit with the shipyard's production technology and retain a wooden hull while having a light, strong and durable boat, a cold-moulded hull has been designed. The final hull is made of three layers of 2.5mm African mahogany (*khaya anthotheca*) veneers, the outer one running fore and aft to conserve a traditional look, and sheathed with an E-glass DB 300 for a see-though finish. Douglas fir (*pseudotsuga menziesii*) longitudinal stiffeners and plywood frames and bulkheads complete the hull shell structure.

The scantlings of the vessels have been checked and proven to comply with the ISO 12215-5 [24] for the hull, also considering the global load case recommended in the ISO 12215-6 [23]. Furthermore, the rudder and centreboard and inherent supporting structures have been demonstrated to meet the requirements of the ISO 12215-8 [25] and ISO 12215-9 [26] respectively.

6.1 (b) Stability

As a sailing vessel of hull length greater than 6m, the Thames A Rater is to comply with the ISO 12217-2 [21]. Although compliance can be demonstrated in this case by a capsize recovery test only, at a design stage and in order to ensure compliance with category C, requirements such as downflooding opening and heights, wind stiffness and flotation requirements have been considered.

6.2 PERFORMANCE

6.2 (a) Sails

Vortex lattice method (VLM) was employed for the design of the sails, with particular attention to minimising the vortices created. Furthermore, to maximise the aerodynamic efficiency of the jib, a flatter foredeck was implemented, thus reducing the gap between the foredeck and foot of the jib to a value much lower than the current A Raters. The effect is a 6% reduction in drag and a 4% improvement in lift [20].

Finally, two reefs were added based on the velocity prediction programme (VPP) detailed in Section 6.2 (c), representing respectively 75% and 45% of the full main sail area. Those particular values have been chosen to give the best performance to the new A Rater in a wider range of weather conditions.

6.2 (b) Balance

Absolutely critical to achieve a performance yacht, balance between the hydrodynamic forces applied at the centre of lateral resistance (CLR) and the aerodynamic forces acting through the centre of effort (CE) of the sails must be achieved. However, good balance does not consist in a perfect equilibrium where no rudder angle is needed; a slight weather helm is preferred. Indeed, a small amount of weather helm makes the yacht safer in gusts as it lifts up into the wind, thus depowering the sails. Furthermore, it provides good feedback to the helmsman. Finally, weather helm means the rudder side force acts together with the centreboard side force.

However, locating the CLR and CE is a complex problem. In this case, the centre of lateral resistance was assessed using the method proposed by Delft [27, 28], while the centre of effort was considered as the geometric centre of area of the sails; a common assumption, mostly valid for small and moderate angles of heel [29]. The relative position of the CE in front of the CLR, known as the lead, is typically expressed as a percentage of the waterline length (Lwl). In this instance, a 7% lead was considered, at the upper end of the range suggested by the literature for fractional sloops [17, 20]; this is justified by the beamy design coupled with a high aspect ratio rig. Furthermore, should the lead prove to be unsuitable during sea trials, it is always easier to move the CE back (by raking the mast for instance), than bringing it forward. Consequently, greater lead values are seen as safer.

6.2 (c) Velocity Prediction Programme

The planing abilities of the Thames A Rater are not particularly well modelled in most commercially available VPP software; hence the creation of a six degrees of freedom VPP with the addition of planing behaviour by adapting Savitsky's planing theory for flat plates [30, 31]. The hydrodynamic resistance was calculated in accordance with the Delft Systematic Yacht Hull Series [32], and the aerodynamic forces are based on the Offshore Racing Congress coefficients [33].

The VPP created was first used to assess the performance of the original *Scamp*, and later compare it with the proposed design. The results in 4, 8, 12 and 16 knots of true wind speed (TWS) are presented in Figure 19.

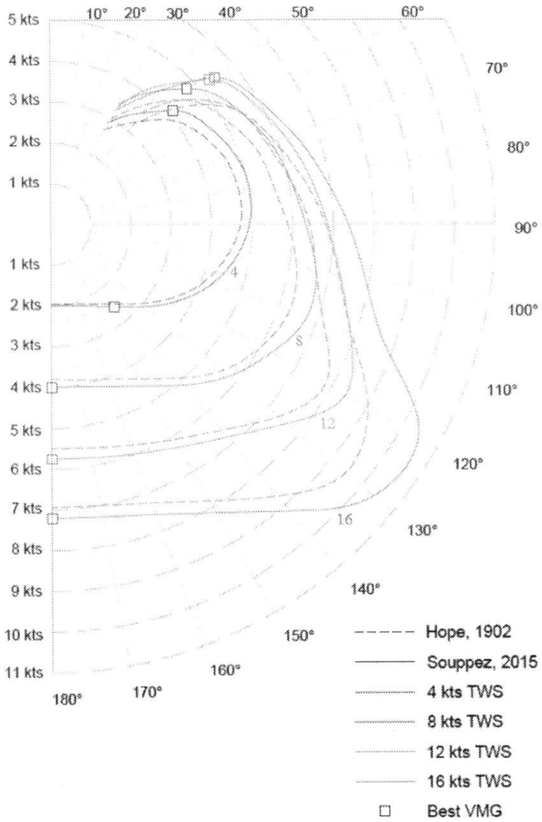

Figure 19: VPP results comparison.

The new design appears to be faster than the original one, both upwind and downwind. Indeed, in addition to the lower hydrodynamic resistance and higher drive force, additional factors contributed to the large increase in performance.

Firstly, the reefed mainsail, higher CLR location, added form stability and more efficient hiking led to a faster boat upwind, especially in stronger winds as the crew can fully handle the yacht. Downwind, the flatter hull shape combined with the ability for the crew to move longitudinally thanks to the new cockpit layout enhances the planing capabilities, that can be seen between 110° and 130° of true wind angle at the higher wind speeds, where the gains in speed are the most significant. Finally, the increase in speed is much smaller dead-downwind, where the stability, longitudinal balance and high aspect ratio foils have very little impact on the sailing.

6.3 CONCLUSIONS

The proposed design has been shown to comply with the requirements for category C for both stability and structure. Furthermore, the VPP created allowed to demonstrate and quantify the benefits of the modern naval architecture techniques employed to improve the performance of the A Rater.

7. CONCLUSIONS

The Thames A Rater class provides a fantastic insight into the history of British inland racing, but also represents an example of a traditional class that remains particularly attractive for racing. The Thames A Raters demonstrate the possibility of preserving their legacy while evolving with technology. The rig and sails being a perfect example, as well as the use of composites and contemporary construction methods.

Based on those observation, the modern design of Thames A Rater was undertaken. The new design is based on an existing linesplan, thus complying with the class rule requirements. However, the application of modern techniques such a computational fluid dynamics and parametric optimisation allowed to significantly improve the hydrodynamics of the hull. Ergonomics and anthropometrics have been considered to provide a cockpit layout more in line with today's racing crew organisation. Throughout the project, the rules and regulations, namely the RCD II and relevant ISO standards, have been applied, with particular emphasis on stability and scantlings. The more efficient sail plan, optimised thanks to vortex lattice method and additional considerations, such as the reduction of the gap between the foot of the jib and foredeck, resulted in a significant increase in performance, highlighted and quantified by the VPP. The new design allows to conserve a traditional appearance, yet making the boat more competitive.

There is therefore a place for modern naval architecture techniques, not only for the conservation of historical crafts, but also to ensure their sustainability through time, by offering additional performance and compliance with modern regulations, while retaining the original spirit of tradition.

8. REFERENCES

1. LAITY, M., 'Thames Clubs', *Classic Boats,* 2001
2. LAITY, M., 'Bell Boy', *Classic Boats,* 2007.
3. FROUDE, W., 'Experiments of the surface friction experienced by a plane moving through water', *Report to the British Association for the Advancement of Science*, 1872.
4. FROUDE, W., 'Report to the Lords Commissioners of the Admiralty on experiments for the determination of the frictional resistance of the water on a surface under various conditions', *Report to the British Association for the Advancement of Science*, 1874.

5. UPPER THAMES SAILING CLUB, 'Upper Thames sailing club gallery', *Available online: http://bew.utsc.org.uk/*, 2015.

6. PHILLIPS, M., 'Ulva', *Available online: http://www.tradboatrally.com*, 2005.

7. THAMES SAILING CLUB, 'The Thames A Rater Association Rules', *Thames Sailing Club*, 2007.

8. BROADS AUTHORITY, 'Norfolk and Suffolk Broads Act', *Available online: http://www.broads-authority.gov.uk/*, 1988.

9. EUROPEAN PARLIAMENT, 'Directive 2013/53/EU', *Official Journal of the European Union*, 2013.

10. KEMP, D., 'Dixon Kemp's manual of yacht and boat sailing and yacht architecture. New and eleventh edition. Edited by B. Heckstall-Smith and Linton Hope', *Horace Cox, London*, 1913.

11. INTERNATIONAL TOWING TANK CONFERENCE, 'ITTC recommended procedures and guidelines: model manufacture, ship models', *23th International Towing Tank Conference*, 2002.

12. KEUNING, J. A. and KATGERT, M., 'A bare hull resistance prediction method derived from the results of the delft systematic yacht hull series extended to higher speeds', *Delft University of Technology*, 2008.

13. REMMLINGER, U., 'Bare hull resistance prediction based on the Delft Systematic Yacht Hull Series', *Ulirch Remmlinger*, 2014.

14. LAUNDER, B. and SPALDING D., 'Mathematical models of turbulence', *Academic Press, London*, 1972.

15. SOUPPEZ, J-B., 'On the use of unstructured meshes for free surface flow around a yacht hull', *The University of Auckland*, 2014.

16. CELIK, I. B., 'Procedure for estimation and reporting of discretization error in CFD applications', *West Virginia University*, Undated.

17. WALLIS, S., 'Marine craft design and development: keel and ballast', *Southampton Solent University*, 2011.

18. SOUPPEZ J-B., 'Design and production of a one-off 32 feet wooden racing yacht dedicated to Southampton Solent University', *Southampton Solent University*, 2013.

19. STEWART, J., 'The design of a Thames A Class Rater', *Southampton Solent University*, 2002.

20. LARSSON, L. and ELIASSON, R., 'Principles of yacht design', *Adlard Coles Nautical, London*, 2007.

21. INTERNATIONAL ORGANISATION FOR STANDARDISATION, 'BS EN ISO 12217-2:2015 - Small Craft – Stability and buoyancy assessment and categorization. Part 2: Sailing boats of hull length greater or equal to 6m', *International Organisation for Standardization*, 2015.

22. LE DEROFF, J. Y. and LACHIKINE, P., 'Righting moment measurement on Laser class', *Ecole Nationale de Voile*, 2001.

23. INTERNATIONAL ORGANISATION FOR STANDARDISATION, 'BS EN ISO 12215-6:2008 - Small Craft – Hull Construction and Scantlings. Part 6: Structural Arrangements and Details', *International Organisation for Standardization*, 2008.

24. INTERNATIONAL ORGANISATION FOR STANDARDISATION, 'BS EN ISO 12215-5:2008+A1:2014 - Small Craft – Hull Construction and Scantlings. Part 5: Design Pressures for Monohulls, Design Stresses, Scantlings Determination', *International Organisation for Standardization*, 2014.

25. INTERNATIONAL ORGANISATION FOR STANDARDISATION, 'BS EN ISO 12215-8:2009 - Small Craft – Hull Construction and Scantlings. Part 8: Rudders', *International Organisation for Standardization*, 2009.

26. INTERNATIONAL ORGANISATION FOR STANDARDISATION, 'BS EN ISO 12215-9:2012 - Small Craft – Hull Construction and Scantlings. Part 9: Sailing Craft Appendages', *International Organisation for Standardization*, 2012.

27. GERRITSMA, J. and KEUNING, J. A., 'Further experiments with keel-hull combinations', *SNAME First Tampa Bay Sailing Yacht Symposium*, 1986.

28. KEUNING, J. A. and VERWERFT, B., 'A new method for prediction of the side force on keel and rudder of a sailing yacht based on the results of the Delft systematic yacht hull series', *The 19th Chesapeake sailing yacht symposium, Annapolis*, 2009.

29. SOUPPEZ J-B., 'Upwind wind tunnel testing of the Stewart 34', *The University of Auckland*, 2014.

30. SAVITSKY, D., 'Hydrodynamic design of planning hulls', *SMANE transactions*, 1964.

31. SAVITSKY, D. and BROWN, W. P., 'Procedures for hydrodynamic evaluation of planning hulls in smooth and rough water', *Marine Technology*, 1976.

32. KEUNING, J. A. and KATGERT, M., 'A bare hull resistance prediction method derived from the results of the delft systematic yacht hull series extended to higher speeds', *Delft University of Technology*, 2008.

33. OFFSHORE RACING CONGRESS, 'ORC VPP Documentation', *Available online: http://www.orc.org*, 2013.

9. AUTHORS BIOGRAPHY

Jean-Baptiste R. G. Souppez holds the current position of Lecturer in Yacht Design and Composite Engineering at Southampton Solent University. He is responsible for delivering a number of units on the BEng (Hons) Yacht and Powercraft Design and BEng (Hons) Yacht Design and Production, and is a member of the ISO working group 18 (ISO 12215). He graduated from the BEng (Hons) in Yacht and Powercraft Design at Southampton Solent University, from the Diploma in Practical Boatbuilding at the International Boatbuilding Training College and from the MEngSt in Yacht Engineering at the University of Auckland.

18 MINUTES – A FORENSIC ANALYSIS OF THE LOSS OF RMS LUSITANIA

I E Winkle, MRINA (Retired), UK

SUMMARY

When Kapitanleutnant Schwieger, the commander of U-20, fired a torpedo into the starboard side of RMS *Lusitania* on 7th May 1915, he set in train a sequence of flooding events leading to the total loss of the ship in only 18 minutes with the loss of 1198 lives. Using the resources of Glasgow University's Archive of original *Lusitania* design drawings, the transcript of the British Wreck Commissioner's Inquiry and other published material, the author reconstructed the vessel's hydrostatics and capacities data to deduce key stages of the flooding process, identifying their underlying cause and those steps that could have been taken to prevent her loss.

In conclusion, if more attention had been paid to the technical details of her loss at the official Inquiry (officially precluded by the Wreck Commissioner), *Lusitania* might have had a similar influence on SOLAS Regulations to that of her ill-fated competitor *Titanic*.

1. INTRODUCTION

The material of this paper was originally prepared for, and presented at, a Lusitania Conference organised for the centenary of her loss at the Clydebank Town Hall on 13th May 2015. Although there have been at least two other attempts at a forensic analysis of the loss by American authors [1, 2], neither presents a sufficiently convincing naval architectural assessment of the stages of the loss which could explain both the timing and mechanism of flooding – other than vague references to flooding through open portholes.

Three major data sources have been used to develop this analysis:

- Glasgow University Archive,
- British Wreck Commissioner's Inquiry Report [3],
- 'Engineering' Article of 2 Aug 1907 [4],

together with general information taken from the Lusitania Resource Web Site [5]

1.1 GLASGOW UNIVERSITY ARCHIVE

Here is the major resource of original drawings including a 1:48 scale lines and body plan [6] which has been used to calculate a full set of hydrostatic data (shown in Table 1) together with a wide range of general arrangement drawings which were used to define boundaries for bunker and boiler room capacity calculations (summarised in Table 2) as well as identifying compartment and ventilation systems critical flooding points. The midship section [7] was particularly useful in identifying the detailed internal structure of the boiler rooms and adjacent spaces as well as the 'as built' shell and deck plating arrangements. Archive copies of Lloyd's Registers also confirmed the 1915 operational draught and displacement.

1.2 WRECK COMMISSIONER'S INQUIRY

The testimony of this five day inquiry is reported [3] in full in numbered paragraphs specific to each of the witnesses called. In particular, the testimony of the Superintendent Engineer, Alexander Galbraith (who confirmed the modification to the vessel since build), Captain William Thomas Turner (who stated the duration of the incident from observation of his pocket watch and the approximate draught at the time of the incident) and his Quartermaster, Hugh R Johnston, (who reported the progression of heel as read from a clinometer on the compass binical) was critical to establishing the initial parameters for the incident. In addition, the three survivors from Boiler Rooms 1 (Madden and Davis) and 2 (McDermott) confirmed the nature and origin of the initial stage flooding as well as the fact that there was no initial boiler explosion.

1.3 'ENGINEERING' 2 AUG 1907

This 57 page article [4] discusses the detailed engineering of *Lusitania* at the time of build and is particularly useful in its description of the boilers and automatic watertight door closure arrangements as well as offering a range of relevant photographs. From this, the permeability of the 22ft. long by 17.5ft. diameter boilers was calculated at 9.5% (based on Figs 93 & 95 [4]) giving an average boiler room permeability of 76.5% to top of boilers and about 95% above.

2. INITIAL CONDITIONS

To establish the sequence of *Lusitania*'s foundering it was necessary to establish both her final loading condition and which compartments flooded. The first is rather more complicated than it seemed initially as there are two quoted draughts and derived displacements for the vessel, namely:

- displacement of 36,690ton (from calculated hydrostatics) at Draught (mld) of 33.5ft., Beam (mld) of 87.5ft. and a Depth (mld) to Shelter

Deck of 60ft. 4.5in., giving a Deadweight of 11359ton (including 7000ton of coal) based on the ship as completed in 1907. Her Wikipedia entry quotes a quite unbelievable 44,060ton Displacement (corresponding to a draught of about 38ft.) for this condition!

- after her loss, the Superintendent Engineer of Cunard Line, Alexander Galbraith, stated at the British Wreck Commissioner's Inquiry, that her Displacement was 41,440ton [3 ¶11] at a Draught (mld?) of 36ft. [3 ¶10], Beam (ext) of 88ft. [3 ¶8] with Depth as above giving a substantial potential increase in Deadweight. This seems to agree with the Lloyd's Register entry for 1912-1915.

Later Galbraith explained [3 ¶21] that **additional** heavy plating had been added from the Shelter Deck down to 14.5ft. below the Upper Deck (i.e. 6in. below the 36ft. waterline) which explains the 88ft. extreme breadth. In addition he explains that the Stringer Plate was doubled and the 12 Watertight Bulkheads had been strengthened "to stand the necessary pressure in the event of an accident". So it seems that a ship designed to operate at a Summer Load Line of 33.5ft. (mld) was now operating at 36ft. (mld?) with the same bulkhead arrangement as built.

Her 25 knot service speed on 23 double-ended and 2 single Scotch boilers had been reduced since Nov 1914 to 21 knots using 6 fewer double-ended boilers by closing down Boiler Room 4. Instead of using 910ton of coal per day to maintain 25 knots this would reduce proportionately to about 682.5ton/day at 21 knots. After 5.88 days in transit (predominantly at 21 knots), from noon departure from New York on May 1st, about 4014ton of coal apart from fresh water and stores would have been consumed, lightening the ship and reducing its draught. In his testimony to the Inquiry, Capt. William Thomas Turner stated [3 ¶202] that the draught at the time of her loss was "about 33ft. 10in. approximately" although no trim was given. Taking all consumables into account, about 1630ton of ballast would have been taken on board to achieve this draught and improve her handling and stability. Assuming this was in the double bottoms tanks forward, this would produce a trim of about 1.325ft. by the stern at the time of the incident.

With Boiler Room (BR) 4 closed down it has been assumed that only those bunkers in way of BRs 1, 2 & 3 would be in service. The capacities of these were proportioned from the 7,000ton overall capacity of the ship amounting to some 5,532ton distributed as shown in Table 2 at a derived bulk coal density of 40.9ft.^{3}/ton. Deducting one third of the 4,014ton consumed from the bunkers around each BR leaves 490ton available around BR1 and only 26ton (2% of capacity) around BR2 while BR3 had over 1,000ton available in its much deeper bunkers. Given the testimony of Frederick Davis (a trimmer) [3, ¶1788] that he had just been to the forward bunker hatch and loaded up his wheelbarrow and returned up the amidships pass in BR1 when the explosion occurred, it can be assumed that the easily won coal from the nearer side bunkers was now exhausted and these were effectively empty (and sealed). Furthermore, BR2 would be starting to draw its coal from the large bunkers beside BR3 (transiting through the dividing bulkhead watertight door) and its side bunkers were effectively empty as all the remaining coal was below the level of the bunker doors and difficult to get at.

3. LOCATING THE DAMAGE

It seems from the *U-20*'s log that Kapitanleutenant Schwieger had manoeuvred to within 700 yards of *Lusitania*'s surprisingly straight track when he fired his Type G torpedo, set to run at a depth of 3m., aiming to hit *Lusitania* abreast of BR4. Having mistakenly estimated *Lusitania*'s speed at 21 rather than the 18 knots that she was actually making, the torpedo struck about 230ft. forward of his target abreast of BR1, having taken some 47 seconds to impact at a speed of 27 knots. There has been much argument concerning the nature of what appear to have been two explosions, the second being far more pronounced than the first. Given that the impact was clearly into a well ventilated, empty coal bunker the second explosion has often been assumed to be due to coal dust, but experimental research in the USA [1, 2] suggests that this is very unlikely. The author's view is that the torpedo fuse may have been faulty and that the initial impact was the torpedo penetrating the vessel's $^3/_4$in. shell with its 160 kg warhead falling into the empty bunker where it exploded shortly afterwards. This explosion, being contained within a relatively small (9,112ft.^{3}) sealed bunker space, appears to have blown out a substantial proportion of the compartment's bulkheads, side shell and deck leaving a gaping hole in the side through which a massive plume of debris laden water and air was ejected to such a height as to knock the third lifeboat (No. 5) on the starboard side out of its falls. Consensus opinion [5] suggests that the impact location was therefore in the after starboard bunker alongside BR1, around frame 210 (just aft of bulkhead 213 separating the bunkers) and about 23.5ft. above keel (just below the Orlop Deck Stringer). The effect was an immediate heel of at least 15 degrees to starboard from the extremely rapid flooding of at least three empty starboard bunkers followed by BRs1&2. These spaces are identified in the forward portion of the Profile and Orlop Deck drawing (Fig 22 [4]) to which the inflow has been added to form Figure 1.

It is clear from the testimony of Thomas Madden [3, ¶1731-1734] that BR1 flooded fairly quickly from the starboard side ("I ran to the watertight door, that was shut down, and by the time I got back it was coming through the boilers. There would be about a foot and a half then") with an additional 3ft. of flooded space below the stokehold flat above the tank top. As he escaped using the port side escape ladder up the Ventilator he "got knocked down by the force of the water, but I got up and went up". Both Frederick Davis and Thomas Madden identify the source of the explosion to be on the starboard side of BR1 although each associated it with the opposite end of the compartment. Another trimmer, Mr McDermott [3, ¶1770], in BR2 stated that the force of the flood "water washed me out through the bottom of the ventilating shaft" – a similar experience to Thomas Madden. It is quite clear therefore that BRs1&2 were flooding rapidly from the starboard side shortly after the explosion (although they did not mention significant heel at this time – although it would have been almost instantaneous). As a result, it is unlikely that boiler pressure could be retained in these spaces for more than a minute or so before the furnaces would have been extinguished - a process that would have produced plumes of steam and soot up the funnels and ventilators. Two thirds of the available steam would have been lost very quickly and these rapidly cooling boilers would have formed a heat sink condensing much of the steam from BR3 unless BRs1&2 steam lines had been rapidly isolated. This does not appear to have happened as the duty engineers (in the Low Pressure Turbine Room on the centreline aft of BR4) did not have any insight into the loss of the two boiler rooms forward – only that the steam pressure rapidly reduced from the normal 195 to about 50psi (discussion following [3, ¶668] with Third Engineer George Little). In addition, it might be assumed that the forward bunker might also have been available for flooding as the coal hatch had been open and in use just before the explosion (see Frederick Davis above). However, testimony from the bridge indicates that the automatic system to close the hydraulically activated Watertight (WT) doors below the Main Deck (also shown in Figure 1) had been activated just before the torpedo struck in response to its wake spotted by the lookouts. It is therefore reasonable to assume that this door would form an effective barrier to the passage of water by the time it rose above the door sill. It is also important to note that in the early morning the Captain had given orders for all non-bridge-controlled watertight doors (primarily on the Main Deck) to be closed forcing passengers up the main stairways to the upper deck.

Soon after, the flood water would also have risen through the damaged Lower Deck above the after bunker of BR1 to flood the starboard side of 3rd class accommodation there. This flooding is illustrated in Figure 1 within bounding transverse bulkheads rising to the Upper Deck. With this as the starting point, the mechanisms and stages through which *Lusitania* flooded can now be examined using the time-base of 18 minutes suggested by the Captain's observations of his watch [3, ¶113].

4. BASIS OF CALCULATIONS

In keeping with all the available data, all calculations were done using **Imperial Units**. Initially, time was ignored until the whole flooding process had been examined and then the process was reverse engineered to estimate flow rates within the observed time. The significant stages of the process were identified from a close examination of the flooded compartments, their interconnections and possible flooding points by superimposing the flooded waterlines (taking sinkage, trim and list into account) over the relevant sections of the vessel. All calculations were conducted using an 'added weight approach' in which the asymmetric flooded volumes were integrated manually and added into a multistage spreadsheet where the output from one stage became the input to the next. In this way, an assumed GM_t could be applied to the initial stage and adjusted until the early stage list corresponded to the observed 15 degree relatively steady state. The subsequent stage GM_t was deduced from the additional weight of flood water distributed to take account of the angle of list using 'wall-sided' assumptions which should give fairly accurate estimates of GZ for the initial 15 to 20 degree cases for such a vessel as *Lusitania*.

One of the key components of these calculations was an estimate of the inflow and accumulation from open portholes in each row along the starboard side as they sank beneath the waterline. Although no clear evidence was given as to the total number of portholes open, it is clear from the Inquiry that passengers ignored the captain's early morning order to close all portholes as this was the primary means of ventilating their cabins and public spaces – the forced ventilation system supplying only warm air. For the purpose of the calculations it was assumed that 8, 14 and 15 portholes were open on the lower, main and upper decks respectively, their inflow rates being estimated in relation to their immersion, and diameter using the same flow rate assumptions as in the earlier American studies [1, 2]. Should there have been fewer open portholes, the effect would be to speed up the initial phase of flooding while slowing the secondary phase. The overall foundering process would remain unchanged.

The results of these calculations are summarized in the following Sections.

5. FLOODING – STAGE 1

After about **20 seconds**, the three starboard bunkers would have filled at an initial flood rate of about 3,000ton/min., to produce an immediate list of about 15 degrees. In this condition Lusitania would take up a slight trim of about 0.146 degrees by the bow and the Lower Deck portholes over BR2 and the after end of BR1 would already be underwater.

About **85 seconds later**, with some 4,750ton of water on board and a flow rate reducing to about 1,900ton/min., the wing bunkers are fully flooded, together with the aft end of the Lower Deck space above BR1 (to starboard) and BRs1&2 are flooded to cover most of their boilers (all of their furnaces extinguished by this time). The list remains at about 15 degrees and trim has increased to 0.930 degrees by the bow. This is illustrated in Figure 2 which is a section through the forward portion of BR1 (extracted from [8]). By now the Lower Deck portholes are well underwater and the waterline in approaching those on the Main Deck above. With the added flood water low down, the GM_t rises from about 1.96ft. before impact to about 3.71ft. at this stage.

6 minutes after impact, with about 9,050ton of water filling the 3 starboard bunkers and most of BRs1&2 to the base of the funnel casings at Main Deck level, the Lower Deck and Main Deck portholes would have been shipping about 116ton/min. over nearly half the length of the ship. Figure 3 (extracted from [9]) illustrates the position of the starboard waterline in relation to the rows of portholes. The GM_t is now about 4.323ft. resulting in a list of 15.4 degrees with a trim of 1.826 degrees by the bow. However, water is now flooding up the starboard aft stairwell shown in Figure 4 (extracted from [10]) between the Lower and Main decks above the aft bunker beside BR1. It then finds its way down the forward stairwell to flood the Lower Deck above the forward portion of BR1. As shown in Figure 4, each of these stairwells was fitted with a watertight door which in *Lusitania*'s role as an auxiliary cruiser would have been closed with all the Bulkhead watertight doors to isolate the Lower Deck from the Main Deck. However, as these lower spaces were 3rd class passenger accommodation it is very unlikely that they would have been sealed so as not to entomb their passengers below the Main Deck. However, with some forethought the Captain could have removed these passengers earlier in the day to the public spaces above when in perceived danger zones and sealed all Main Deck stairwell doors. Had this been done **and** the portholes all been kept shut as well, the ship would have survived in this condition, proving the original naval design case to prevent foundering after such side damage. It is unlikely that the function of these doors was understood by the relatively inexperienced crew.

6. FLOODING - STAGE 2

Flooding through portholes and up the stairwells onto the Main deck above BR1 continues so that by about **13 minutes** from impact the larger Upper Deck portholes are now shipping water to add to the total rate of about 205ton/min.. This is illustrated in Figure 5 [9]. As a result there is now about 10,400ton of floodwater on board, high enough to start reducing the GM_t to about 4.09ft., increasing the list to about 21.0 degrees and the trim to about 2.16 degrees by the bow. At this point another critical flooding point is reached as the internal floodwater rises up through the open stairwell on the starboard side of BR1 boiler casing seen in Figure 6 (extracted from [11]) to start flooding the Upper Deck where it can flow freely all the way through the 3rd Class Dining Room towards the bow. This is the only stairwell (about 24ft.^{2} in plan) to the Upper Deck not to be positioned on the centerline to minimize the risk of flooding while listing, a fact occasioned by the double step placed in bulkhead 223-233 at the forward end of BR1 (due to its excessive length required to accommodate an additional row of boilers). Shortly after, as more water builds up, it can also flow down the centerline stairwell to flood the lower compartments forward of bulkhead 223, effectively destroying the watertight integrity of that bulkhead. This process will continue to accelerate until the Final Stage is reached at about **16 minutes** after impact.

7. FLOODING - FINAL STAGE

At this Stage, with some 11,545ton of floodwater aboard, the GM_t is 3.57ft., the list has risen to 26.9 degrees with a trim of 2.69 degrees by the bow and water is rushing onto the Upper Deck from below at a rate of 250-300ton/min. while the portholes continue to flood at about 350ton/min.. By now the starboard side of the Shelter Deck is underwater, as shown in Figure 7A [9], and the **Final Stage** commences as the bulwark of this deck near amidships passes underwater forming a long weir over which a massive amount of water is free to flood down the forward portion of this deck without hindrance into all forward compartments.

Within a further 45 seconds this flow will reach about 1,500ton/min. and within another 45 seconds it could have reached about 5000ton/min. cascading unchecked through an opening about 100ft.^{2} in section into the forward spaces. At this point the ship is still stable (having taken on some 17,500ton of floodwater) with a GMt of 3.08ft., but with a list of 33.6 degrees and a trim of 4.63 degrees by the bow. Both the bow and the starboard boat deck are now underwater as shown in Figure 7B [9]. Within another 30 seconds the list has reached about 40 degrees and the bridge wing is now underwater – the point at which the Master and Quartermaster both slide down the deck and into the water to give our time of 18 minutes from impact. By now most forward spaces are full of floodwater and the bow plunges towards the sea bed some 300ft. below. The impact with the seabed, while still listing to starboard and trimmed at about 37.5 degrees with the stern well clear of the water, rolls the vessel further to starboard so the funnel and ventilators of BR3 are at the water surface. At this point two passengers floating in the water nearby are sucked into Funnel No. 3 and seconds later expelled again (to live to tell the tale) in a cloud of steam and soot as the vessel rolls back to port somewhat. This would seem to correspond to the point at which the furnaces in BR3 were suddenly extinguished by the massive inrush of water, possibly aided by a fractured steam line as the boilers rolled in their cradles. From this point the after end of *Lusitania* rapidly fills with water and she settles onto her starboard side on the sea bed, so completely obscuring the evidence of the original damage to frustrate posterity.

8. SUMMARY OF FLOODING PROCESS

Figure 8A shows a timeline of the process discussed in Sections 4 to 6 above in terms of variations in GM_t, the list, trim and accumulated floodwater. The nature of the relatively constant list in the first half of the event is clearly seen as is the very slow rate of flooding in Stage

2 between about 6 and 16 minutes from impact which separates the very rapid flooding processes during Stages 1 and 3. The rapid transition in the Final Stage must have come as a major shock to those on board who may have thought they had more time to evacuate over an extended Stage 2. This becomes clearer in Figure 8B which looks at the 3 separate sources of flood water and clearly identifies the impact of the porthole flooding on what would have otherwise been a survivable situation if progressive internal flooding had not been initiated by the open portholes and enhanced by the open stairwell watertight doors.

9. WHAT MISTAKES WERE MADE?

Although the order had been given early in the morning of the disaster to close all watertight doors and portholes, it is unlikely that the critical stairwell doors between the Lower and Main Decks over the coal bunkers were included in this action as there were passengers in their cabins on the Lower Deck for whom these formed the only access to the public spaces above. Indeed, no direct reference to these doors occurred throughout the Inquiry. However, it would have been possible to move all passengers around breakfast time to the public spaces above and seal these Lower deck spaces until shortly before arrival in Liverpool.

Furthermore, no attempt appears to have been made to follow up on the order to close all portholes and deadlights. It seems fairly straightforward to have set up a patrol of the passenger spaces with a few stewards to ensure that this order was not negated by the actions of passengers wanting some fresh air on what was a fairly warm spring day. The potential impact of open portholes on the flooding of a vessel listing after a torpedo attack does not seem to have been appreciated even during the Inquiry despite a few passing references to water flooding through portholes from some of the surviving passengers.

Had both these actions been taken, it is clear that *Lusitania* would have remained afloat, albeit with significant list and trim. If a means had been found to flood the port bunkers by isolating the steam supply from BR3 and directing it to power the hydraulics which operated the relevant bunker doors then the list could have been removed to assist in launching all lifeboats under less panicked conditions. This course of action had occurred to Captain Turner, but all power had been lost by then, and there had been no time for a detailed survey of the actual flood damage to know how to react.

It is even conceivable that with power from BR3 isolated and flooding contained, it might have been possible to limp towards the safety of Cobh Harbour.

10. LESSONS *NOT* LEARNED!

The Commissioner, Lord Mersey, presiding over *Lusitania*'s short 5 day Inquiry [3], after suffering the long drawn out *Titanic* Inquiry only 3 years earlier, considered that he had already addressed all significant technical problems relating to the safety of passenger ships and steadfastly refused to allow questioning that attempted to draw out any such lines of enquiry. The result was that a number of lessons were not learned that could have contributed to the early development of SOLAS Regulations. Such major contributions might have included:

10.1 OPEN PORTHOLES

the dangers of open portholes/deadlights, particularly when associated with substantial longitudinal subdivision. These would always prove a problem in vessels without controllable air conditioning as was the case only a year later in the loss of *Britannic* while on duty as a hospital ship, where the high temperatures in the Aegean had encouraged the nursing staff to open the portholes for ventilation which contributed to the final stage of her loss.

10.2 CROSS-FLOODING

if wing compartments are to be included in a design (considered essential by the Royal Navy for *Lusitania*'s potential role as an auxiliary cruiser) then there should be some mechanism to establish rapid cross-flooding between opposite spaces to minimize the risk of substantial list developing. It was not until the loss of *Andria Doria* in 1956 that this problem was finally recognized, when the inability to cross-flood empty wing fuel bunkers after a collision had opened up only one side, led to her eventual capsize. This concept was not adopted into Regulations until SOLAS 60.

10.3 LIMITING LIST FOR LIFEBOATS

the importance of limiting list in damaged conditions to allow the safe launch of lifeboats. At 15 degrees plus, in the case of Lusitania, no port side lifeboats were successfully launched, several being overturned as they rubbed down the side of the vessel, catching on the in/out riveted seams as they descended. On the starboard side the boats swung out and passengers had to jump to enter them (not always successfully). Not until SOLAS 60 was a limitation of 7 degrees imposed on the damage stability of passenger ships, later amended to allow up to 12 degrees for '2 compartment vessels' in 1990.

10.4 INDEPENDENT POWER SOURCES

the importance of ensuring an independent power source for auxiliary systems. It is unfortunate that all four boiler rooms on *Lusitania* seem to have supplied a single steam chest in the engine room which fed both main and auxiliary machinery. A similar fate befell the *Costa Concordia* when its two dedicated diesel-electric generator rooms were both flooded and all control of the ship and its auxiliaries was lost except for its small emergency generator which only functioned for emergency lighting and communications.

11. CONCLUSIONS

Even Kapitanleutenant Schwieger aboard U20 was surprised at the impact of his single torpedo strike on *Lusitania*. While the Navy, reeling from their own contemporary losses to submarine launched torpedoes, seemed desperate to pin the blame for the loss of life on Captain Turner for not following Naval Orders (encouraging passenger testimony about two or more observed torpedoes, despite the evidence of the lookouts and senior officers that there was only one track, but two 'concussions', 'one immediately after the first' [3, ¶107]), their own man on board, Reserve Captain Anderson, to whom the order was given to ensure all bulkhead watertight doors were closed [3, ¶197-198] does not seem to have appreciated the potential impact of the stairwell watertight doors on the watertight integrity of the damaged ship. Against this background and many volumes of text recounting her loss, there has been no serious attempt to piece together the actual foundering process to explain what happened and see what might be learned. The two American attempts at forensic analysis [1,2], have many pages of computer output attempting to describe the impact of flooding on the vessel, but neither of them contain a step by step analysis of the flooding process or try to draw any useful conclusions. The author hopes that he has been able to overcome this lack of detailed technical appraisal within this paper and that over an hundred years after the event, it is finally clear what caused this avoidable disaster.

12. ACKNOWLEGEMENTS

This research was undertaken over the year preceding the Centenary of *Lusitania*'s loss which was commemorated at a day conference at Clydebank Town Hall, a short distance from where she was constructed within the John Brown shipyard, on 13th May 2015. This conference was organised by the Clydeside Maritime Heritage Group, of which the author was a member, and although none of the proceedings were published at the time, it was always intended that opportunities to do so would be sought in the future.

The research relied heavily on the extensive collection of original *Lusitania* drawings held in the University of Glasgow Archive collection (as part of the National Records of Scotland and University of Glasgow Archive Services, John Brown & Co Ltd collection, GB248) and the author is indebted to the staff of that Department for their help and forbearance in dealing with the many requests for the oversized drawings that form the bulk of that collection and for permission to publish those elements referenced [6 to 11].

13. REFERENCES

1. Garzke, W. H. Jr.; Brown, D. K.; Sandiford D.; Woodward, J.; Hsu, P.K., 'The *Titanic* and *Lusitania*: a Final Forensic Analysis', Marine Technology Vol 33, No 4, Oct 1996

2. Garzke, W. H. Jr.; Dulin R. O. Jr.;Hsu, P. K.; Powell, B.; Bemis, F. G., 'The Saga of *RMS Lusitania*', The New York Metropolitan Sections of SNAME and ASNE, Jan 1998

3. 'British Wreck Commisioner's Inquiry', http://www.titanicinquiry.org/Lusitania, Jun 1915

4. 'The Cunard Turbine-driven Quadruple-screw Atlantic Liner Lusitania', Engineering, 2 Aug 1907

5. 'The Lusitania Resource: Lusitania Passenger & Crew, Facts & History', www.rmslusitania.info

References 6 to 11 below are from the National Records of Scotland and University of Glasgow Archive Services, John Brown & Co Ltd collection, GB248:

6. 'Lusitania, 1/4" Scale Lines & Body Plan', UCS 1/110/367/60

7. 'Lusitania, 1/8" Scale Midship Section', UCS 1/110/367/42

8. 'Lusitania, 1/8" Scale Sections', UCS 1/110/367/41

9. 'Lusitania, 1/4" Scale Rigging', UCS 1/110/367/21

10. 'Lusitania, 1/8" Scale Main and Lower Decks', UCS 1/110/367/40

11. 'Lusitania, 1/8" Scale Upper and Shelter Decks', UCS 1/110/367/39

14. AUTHOR'S BIOGRPHY

Ian Winkle retired from his post of Senior Lecturer at the Universities of Glasgow and Strathclyde in 2005 having been involved in naval architecture, ship design and ship production subjects since 1979. He has maintained a special interest in ship losses, particularly those involving capsize, since the *Herald of Free Enterprise* tragedy in 1987. For the last four years he has been a volunteer guide and researcher at the Glasgow Riverside Museum where he was encouraged to pursue historical research. He now lives in Yeovil.

							Galbraith 41440ton				Captain Turner 33ft 10in
T mld (ft)	**8**	**12**	**16**	**20**	**24**	**28**	**32**	**36**	**40**	**44**	**48**	**33.833**
Mld Vol (ft3)	224227	375534	543878	719506	895097	1076076	1259689	1446693	1635636	1827885	2021192	1345383
Mld Disp (t)	6406	10730	15539	20557	25574	30745	35991	41334	46732	52225	57748	38440
VCB (ft)	4.531	6.753	9.003	11.188	13.311	15.446	17.570	19.692	21.810	23.930	26.046	18.542
LCB (ft)	1.573	1.407	-2.973	-5.700	-7.090	-8.706	-10.571	-12.252	-13.951	-15.452	-16.906	-11.341
LCF (ft)	2.140	-0.011	-3.479	-9.744	-14.842	-19.264	-22.653	-25.306	-27.712	-29.369	-31.437	-23.869
BML (ft)	3522.161	2517.738	1956.104	1620.094	1410.993	1236.816	1099.913	992.518	909.435	845.152	799.683	1050.699
KML (ft)	3526.693	2524.491	1965.107	1631.282	1424.304	1252.262	1117.483	1012.210	931.245	869.082	825.729	1069.241
BMt (ft)	62.194	44.290	33.774	26.842	22.683	19.451	16.964	15.027	13.604	12.175	11.127	16.077
KMt (ft)	66.725	51.043	42.777	38.031	35.994	34.896	34.534	34.719	35.414	36.106	37.174	34.619
TPI (t)	85.75	93.91	99.26	102.87	106.32	108.60	110.34	111.86	113.54	114.67	116.29	111.04
MCTI (t-ft)	2474.2	2962.1	3333.0	3651.8	3956.7	4169.5	4340.7	4498.3	4660.1	4839.7	5063.6	4412.9
Cb	0.4215	0.4706	0.5112	0.5410	0.5608	0.5779	0.5920	0.6043	0.6149	0.6247	0.6332	0.5976
Cw	0.5416	0.5931	0.6269	0.6497	0.6715	0.6859	0.6969	0.7065	0.7171	0.7242	0.7344	0.7013

Table 1: Lusitania - Tabular Hydrostatics (Salt Water, including shaft bossings and propellers)
(longitudinal Centroids, positive forward of amiships)

Bunker Capacities

Frames		Vol (P)	Vert M	Long M	Trans M	Vol (S)	Vert M	Long M	Trans M
213-233	BR 1F	15046	326076	2840234	366943	15046	326076	2840234	366943
197-213	BR 1A	9112	228744	1280351	288782	9112	228744	1280351	288782
182-197	BR 2F	12266	278574	1267716	412503	12786	285084	1312176	430218
167-182	BR 2A	15375	328363	970066	538838	15375	328363	970066	538838
137-167	BR 3	48171	1266857	162030		47514	1259072	158416	
107-137	BR 4	29797	635521	-2235662		30312	641665	-2274030	
Total Vol		**129768**	**3064135**	**4284735**		**130145**	**3069004**	**4287214**	
Total Bunkers		**286350**	**6623987**	**14776459**					
			23.133	**51.603**					
			VCG	**LCG**					
				Flooded Bunkers		36944	839904	5432762	1085943
				(W & Centroids)		**1056**	**22.734**	**147.053**	**29.394**
				Plus Forward Bunker		64724	1339860	11944863	1085943
				(W & Centroids)		**1849**	**20.701**	**184.552**	**16.778**

Coal @ 40.9 cu ft/ton

Frames		W	Vert M	Long M	Trans M	W (S)	Vert M	Long M	Trans M
233-245	Fwd Bunk	646	12001	151699					
		W (P)	**Vert M**	**Long M**	**Trans M**	**W (S)**	**Vert M**	**Long M**	**Trans M**
213-233	BR 1F	368	7973	69443	8972	368	7973	69443	8972
197-213	BR 1A	223	5593	31304	7061	223	5593	31304	7061
182-197	BR 2F	300	6811	30996	10086	313	6970	32083	10519
167-182	BR 2A	376	8028	23718	13175	376	8028	23718	13175
137-167	BR 3	1178	30974	3962		1162	30784	3873	
	Total	**5532**	**130729**	**471544**					
		Available	**Used?**	**Left?**					
	BR 1	1828	1338	490	4 wings empty and 156 t used from Cross Bunker				
	BR 2	1364	1338	26	(less than 35.6 ton below 8 ft.)				
	BR 3	2339	1338	1001					
	Total	**5532**	**4014**	**1518**	(2.22 days supply left)				

Forward Boiler Rooms

Frames		Vol	Vert M	Long M	Vol	Vert M	Long M
			to Lower Deck			**to Main Deck**	
233-245	Fwd Bunk	27780	499956	6512101			
233-245	Bunk Coal	26437	490849	6204509			
213-233	BR 1F	61175	1291645	11791362	70775	1671588	13587098
197-213	BR 1A	70445	1370812	10063031	81965	1817413	11725876
182-197	BR 2F	65307	1278081	6761487	75845	1689579	7869461
167-182	BR 2A	64570	1249577	4102487	75108	1646899	4774937
Total Vol inc Bunker		289276	5690071	39230467	331473	7325434	44469474
Total Vol Ex Bunker		261496	5190115	32718366	303693	6825478	37957373
		7471			8677		
		W	**VCG**	**LCG**	**W**	**VCG**	**LCG**
Perm = 0.76/0.95		**5678**	**19.848**	**125.120**	**6824**	**22.475**	**124.986**

Third Class Accommodation on Lower Deck, Perm = 0.95

Frames		Vol	Vert M	Long M	Trans M
213-233	BR 1F	5347	211566	985782	124599
197-213	BR 1A	6701	262481	941308	174902
Weight & Centroids		**327**	**39.346**	**159.950**	**24.859**
182-197	BR 2F	7575	288531	772464	210298
167-182	BR 2A	8215	314194	518590	236586
Weight & Centroids		**429**	**38.171**	**81.763**	**28.301**

Table 2: Capacities (ft.^{3} and ton) of Primary Flooded Spaces and Coal Bunkers

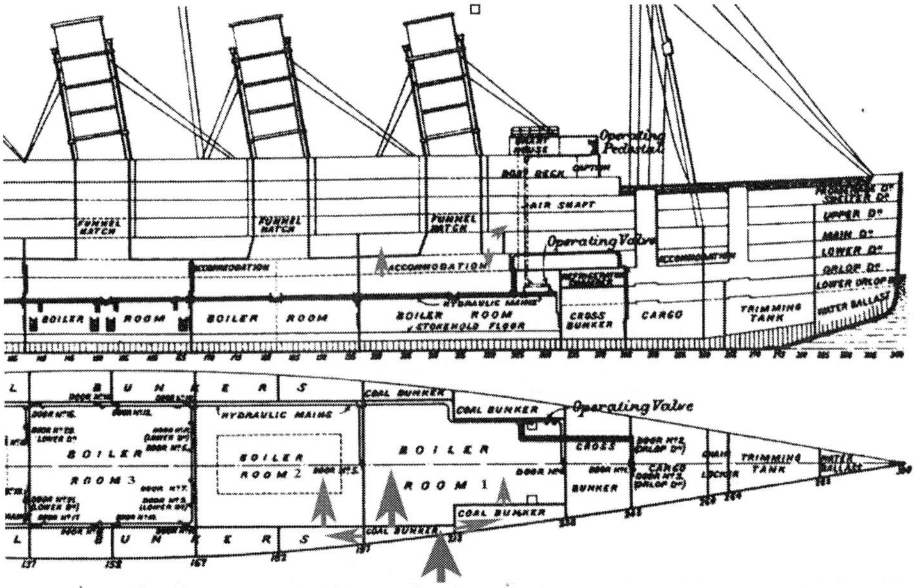

Figure 1: Profile & Deck [4] showing water flow and hydraulic WT door closure system below Lower Deck

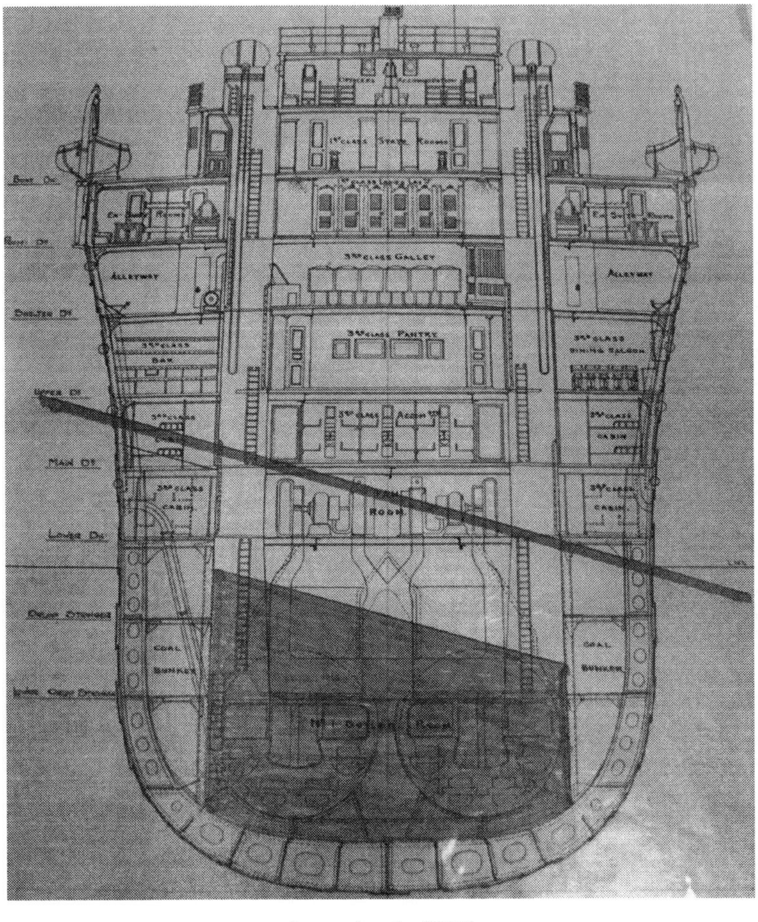

Figure 2: First Stage Flooding after 1.75 minutes – frame 225 [8] at forward end of BR1

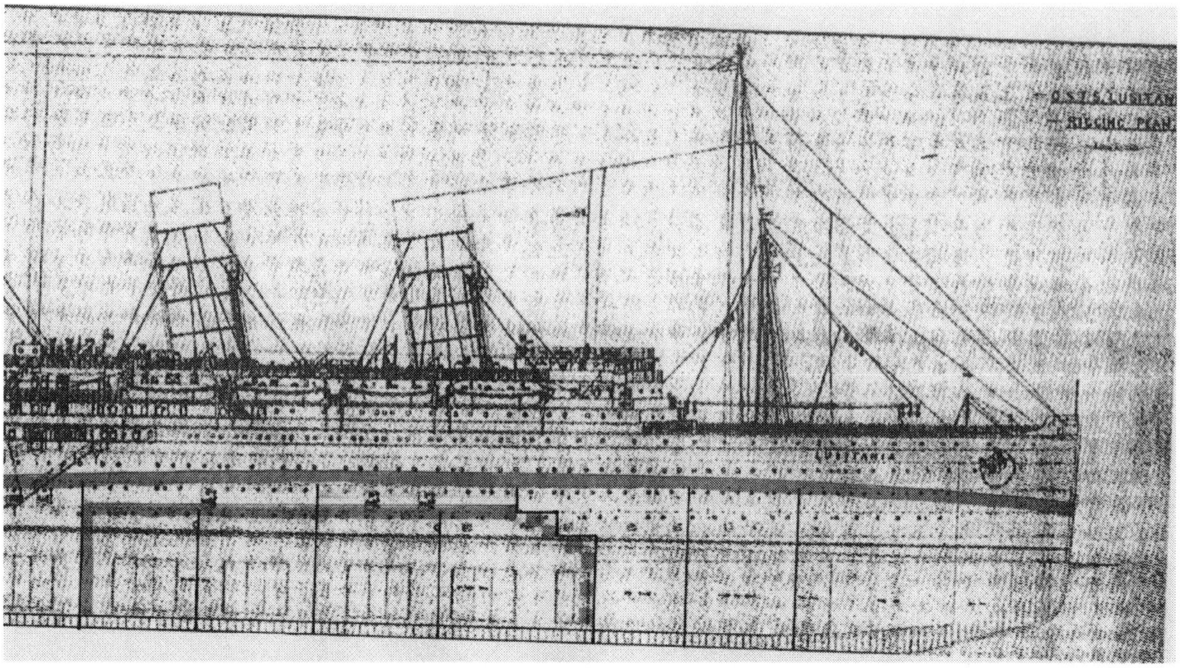

Figure 3: Starboard Waterline and extent of main flooding at end of First Stage Flooding after 6 minutes [9]

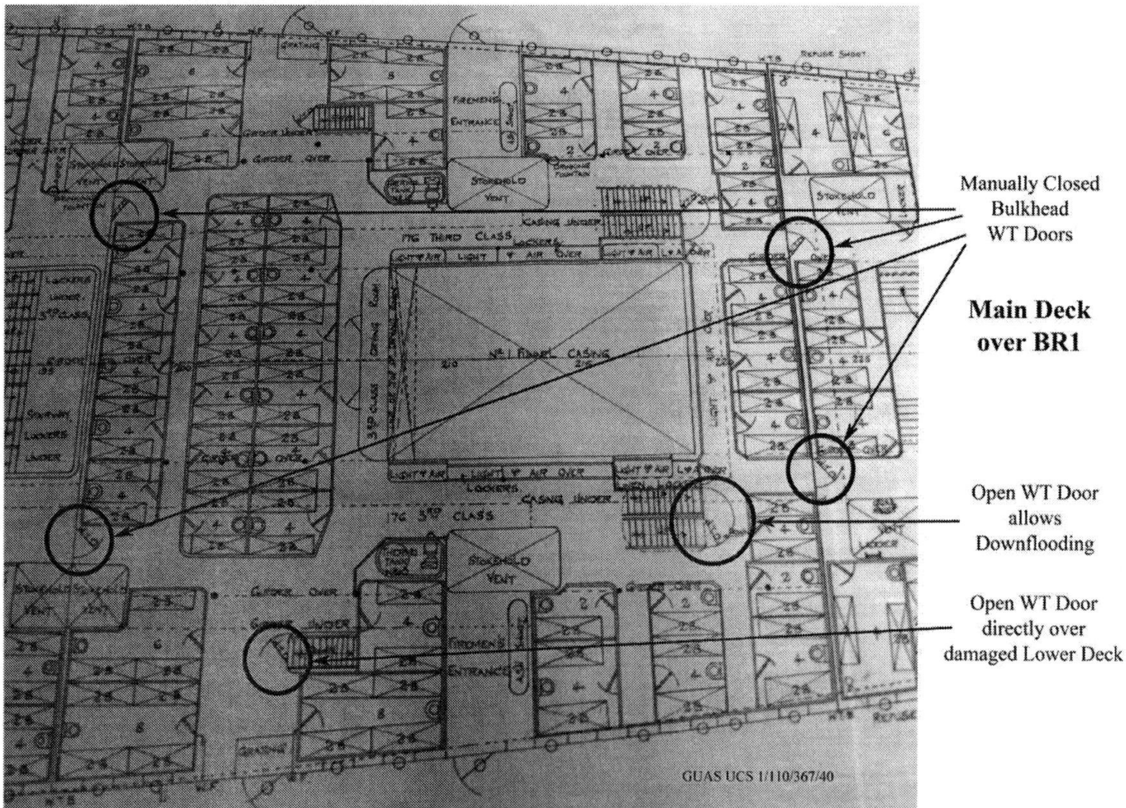

Figure 4: Internal flooding via open WT doors leading to Main Deck flooding over BR1 – end of First Stage, 6 mins [10]

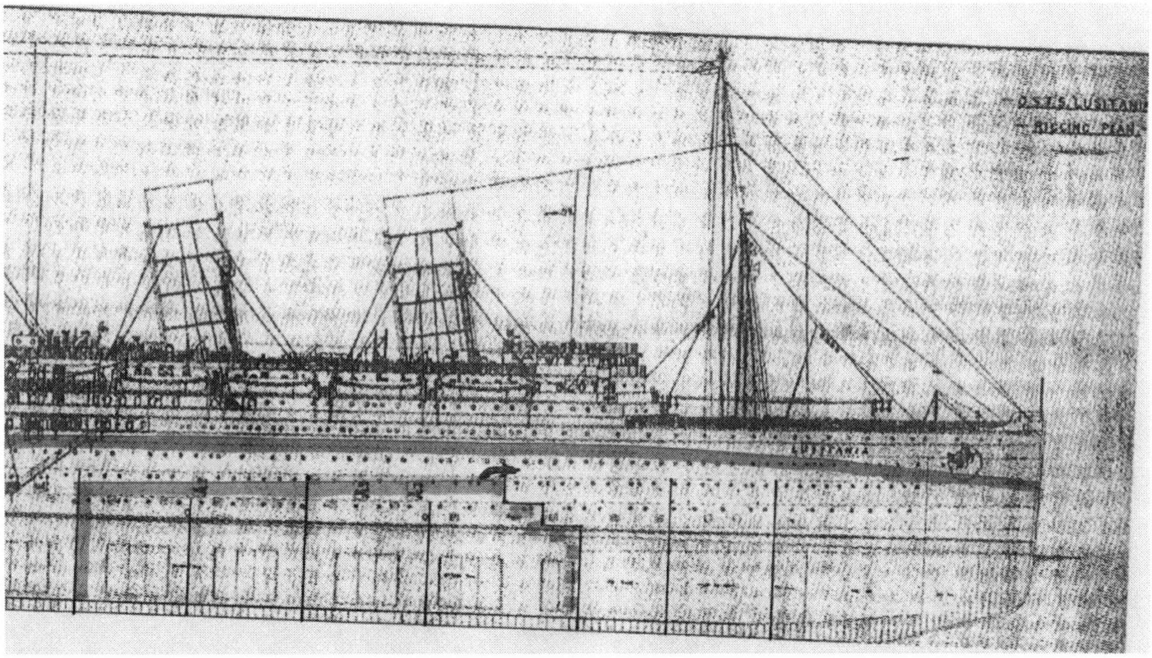

Figure 5: Starboard Waterline and extent of main flooding during Second Stage Flooding after 13 minutes [9]

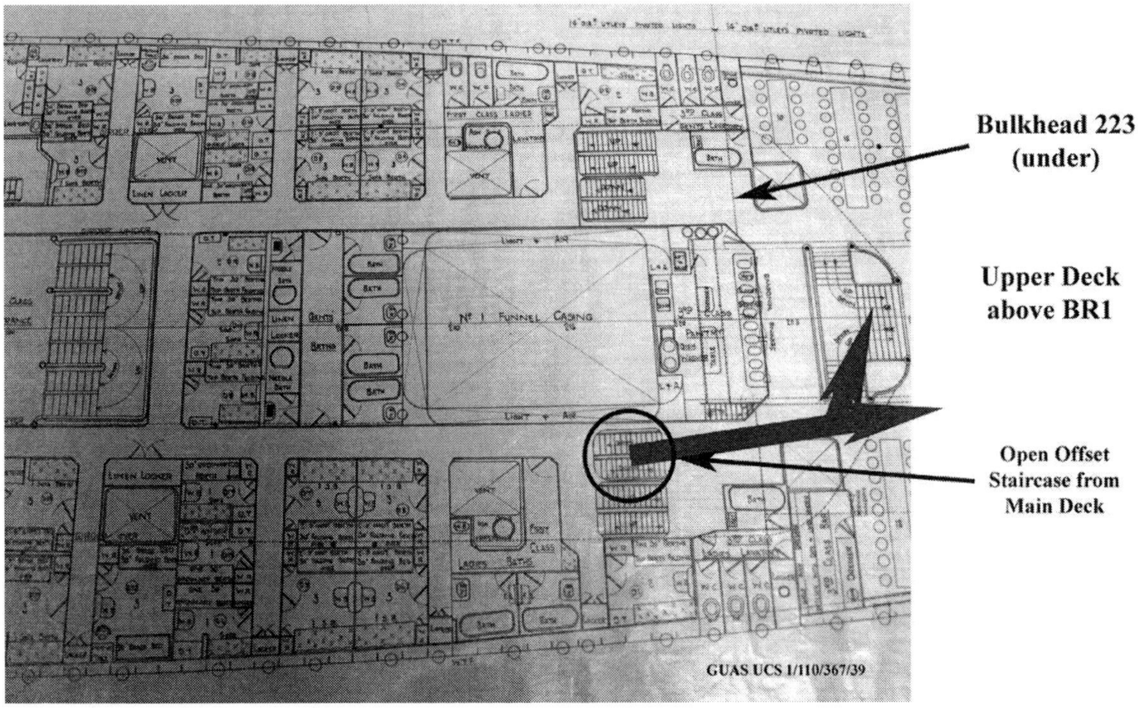

Figure 6: Progressive flooding up starboard stairwell to Upper Deck over BR1 – starts after 13 mins [11]

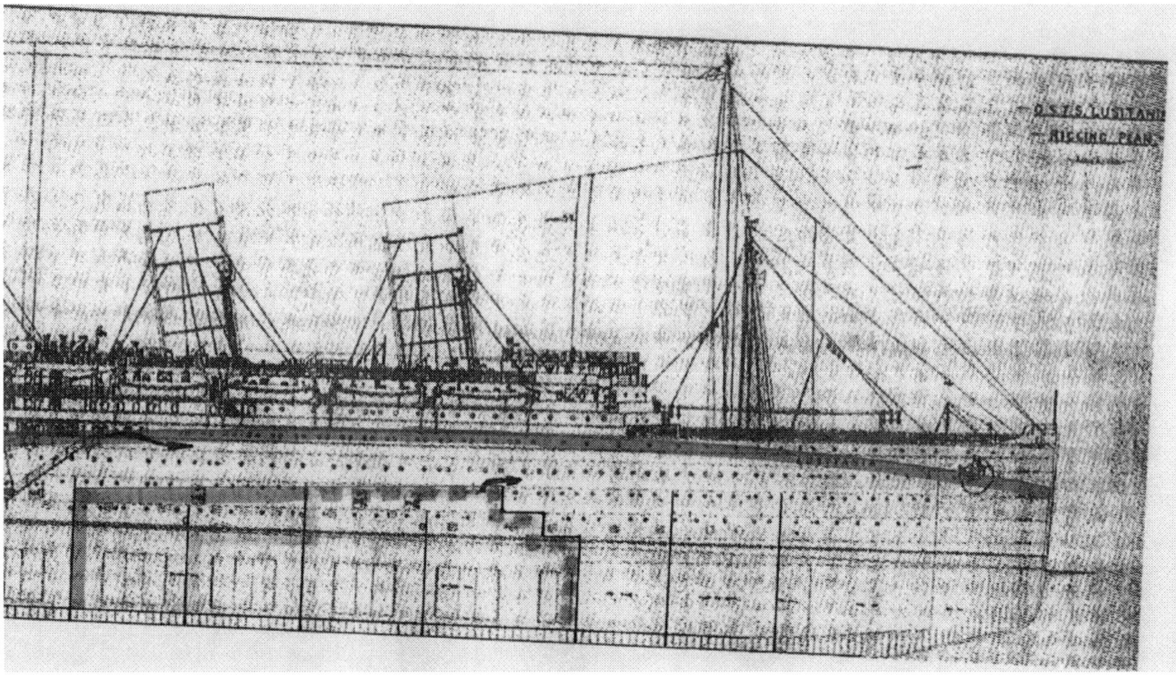

Figure 7A: Starboard Waterline and main flooding points at start of Final Stage Flooding - after 16 minutes [9]

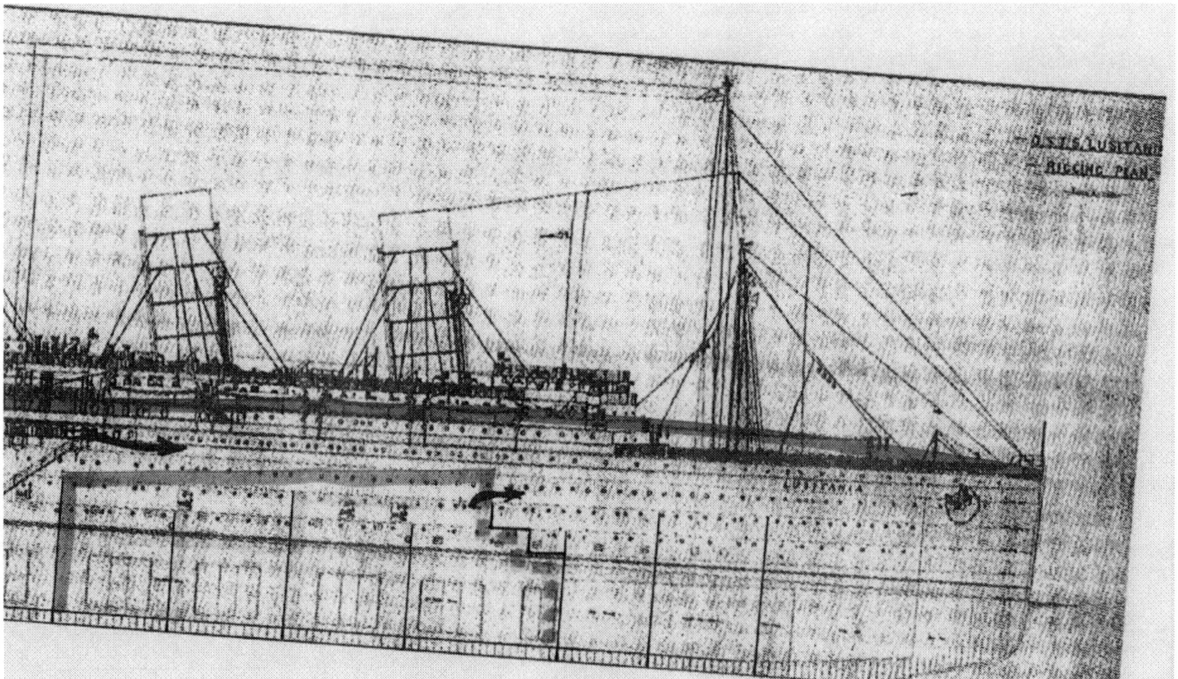

Figure 7B: Starboard Waterline and main flooding points at start of Final Plunge - after 17.5 minutes [9]

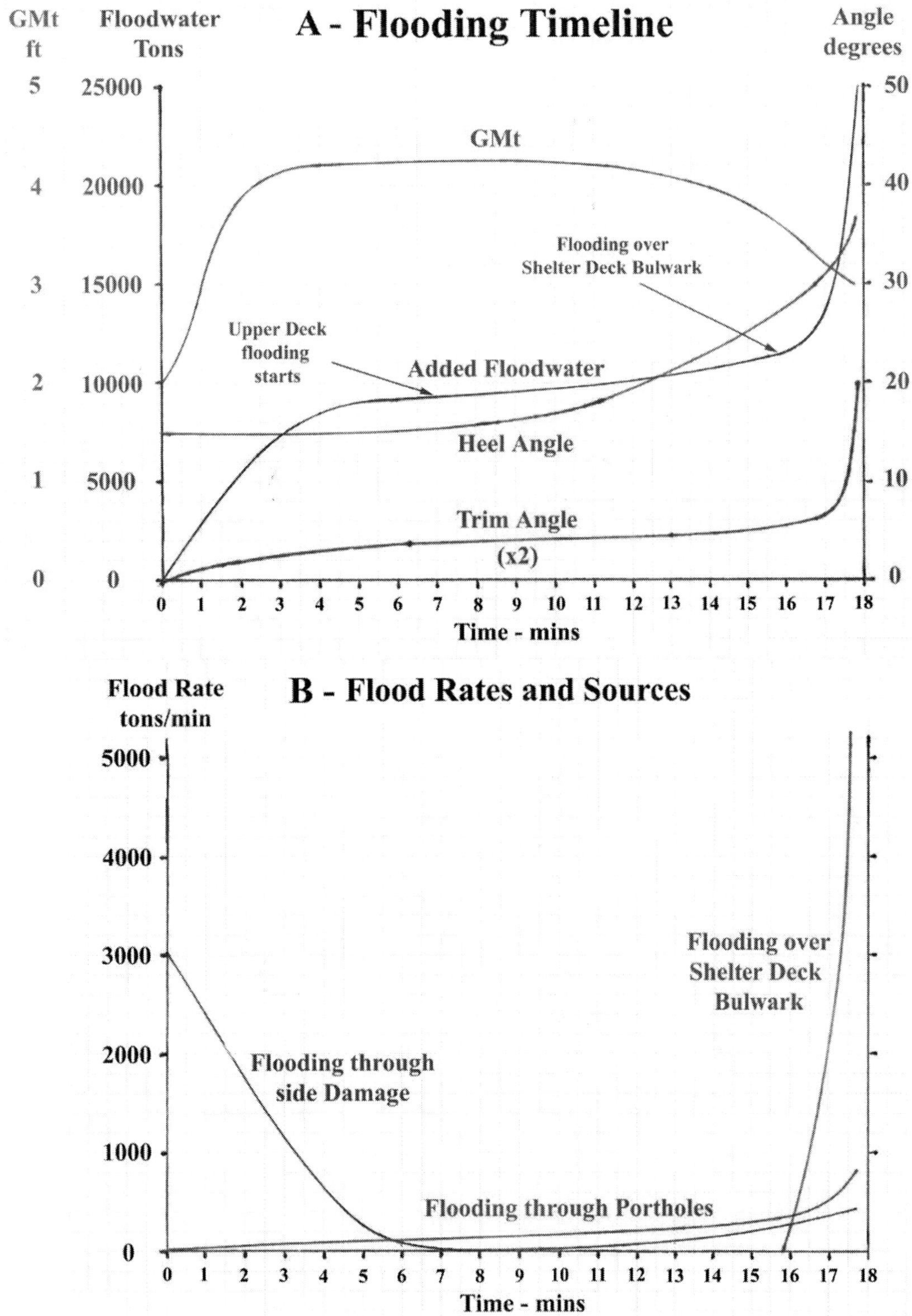

Figure 8: Summary of Flooding Process

DIGITAL MODELLING IN THE EVALUATION OF THE SINKING OF THE HMS VICTORY (1737)

C. E. H. Tonry, P. Mason, M. Patel, and **C. Bailey** Computational Mechanics and Reliability Group (CMRG), University of Greenwich, UK
N. Ball and **S. Stephens**, National Maritime Museum, UK
W. Davies, Wynn Davies 5201 Consultancy Limited, UK

SUMMARY

This paper details a methodology to undertake a forensic investigation into a tragic event – the sinking of HMS Victory 1737. The methodology is an adaptation of Experimental Archaeology (EA). In EA, researchers seek to understand long-lost experiences by sourcing materials, designing, building and finally using a replica of an ancient artefact. By replacing this process with a computer simulation, it is possible to model an artefact which is too large and expensive to consider making in the flesh, and then to deploy it in conditions which are known to be hazardous. Also, it is easy to model all sorts of variations when there is uncertainly around the original design and its operating environment. This is a test for sensitivity.

1. INTRODUCTION

This work builds previous work done by University of Greenwich (UoG) on modelling the structural behaviour of the Medway Queen [1]. This level of computer modelling of an historic ship attracted the attention of the Royal Museums Greenwich (RMG) who were interested in the concept of using modern naval architecture, engineering, material science, and computational methods to understand how historic ships worked, and what the people who designed and built them knew.

Both UoG and RMG are now cooperating on an investigation into a particular set of events in the 18C, and these are summarised as:

- Phase 1 – forensic investigation of sinking of Victory 1737 in 1744
- Phase 2 – what design developments had come into play in the second half of the 18C, and so how would Victory 1765 performed in the same storm?
- Phase 3 – and how would the vaunted French ships of the period have coped when buoyed up (or burdened) with all the maths and early "science" their designers employed?

The 18C was the Age of Enlightenment; the scientific method had been codified and there was an excitement and buzz in people who realised that the knowledge base for design could be improved by use of the scientific method, with its associated mathematical methods. But, there were also others who were suspicious and nervous of new theories.

This paper reports on progress on phase 1 of this investigation.

2. HISTORY OF HMS VICTORY 1737

The first rate three-decker Victory was designed by John Naish in 1725 to the 1719 establishment rules. These rules were a set of dimensions and timber scantlings set out by the Navy Board for each rate of ship.

Victory was built and launched in Portsmouth in 1737, and became the flagship of the Channel fleet in 1741, being the largest ship in the Royal Navy at the time.

There is some uncertainty as to the source of the timbers used to construct this ship – was it from primary sources, or were some of the timbers sourced from the preceding Victory which was broken up in 1720. As well as some ambiguity about what timber was used, there was also criticism of her design:

- French shipwright Blaise Ollivier wrote *"her capacity is very great, yet her upper works are scarce suitable for her lower body, for she is deep-waisted with much sheer"*.
- On 14th April 1740, Jacob Acworth, the Surveyor of the Navy who was in charge of the Navy Board, also expressed some doubts about Victory's design in a letter to her builder Joseph Allin: *Sir John Norris [who was to become Victory's first captain] "has promised to take the ship as she is, but complains much of her height abaft, treble balconies etc, which I was much surprised to hear of. Our ships were too heavy, too loose without these additional encumbrances, which I am sure cannot add beauty, but must be in every respect disagreeable."*

In July 1744, Admiral Sir John Balchen, in command of the Victory, lead a squadron of English and Dutch warships to break the French blockade of the River Tagus at Lisbon in Portugal. After successfully breaking the blockade, Balchen pursued the French vessels and blockaded them in the Spanish port of Cadiz. In September, running low on supplies, he ordered his squadron to return to England.

By October 1744, Victory and the rest of the squadron were homeward bound. As they entered the Western

Approaches of the English Channel on 4th October, they were caught in a violent storm.

During strong gales, the crew of the Augusta, a 60-gun ship in Balchen's squadron, arduously took in sail, as the enormous waves smashed out a great part of the wardroom and starboard gallery. At 10pm that night the logbook of Augusta ominously stated *"lost sight of the admiral and fleet"*. Augusta and the rest of the squadron limped back to Spithead, but the logbook entry was the last recorded sighting of the Victory. She was lost with all hands.

Since the loss, where more than 1000 men died, many have blamed the alleged poor design of the ship. Historians have said that the fleet was not kept in *"a reasonable state of efficiency"* and the early 18th century continues to be described as a period of naval 'complacency' and 'stagnation'. This idea has been challenged recently and the effect on naval shipbuilding has been the subject of some debate. However, much of historical opinion of shipbuilding in this period seems to be largely based on anecdotal evidence - despite Daniel Baugh, as early as 1977, pointing out that: *"It would be satisfying to know whether this tendency of eighteenth-century opinion would stand up in the light of a retrospective engineering analysis."*

The wreck of Victory was discovered in 2008 but the alleged poor design and construction of Victory has been argued in recent research, but without recourse to scientific analysis. The material in the National Maritime Museum, Greenwich, includes original plans, models, logbooks and detailed Establishments. Using modern scientific analysis this material has the potential to answer many of the unanswered questions about Victory's loss.

3. ROLE OF MATHEMATICAL MODELLING AND COMPUTER SIMULATION

There seem to be five possible causes of the sinking in the storm, namely:

1. Structural failure due to an inherent design fault
2. Inundation through the lowest stern window
3. Structural failure due to use of inferior materials
4. An extreme freak wave
5. Seamanship error

The first two can be modelled using conventional engineering techniques which is the focus of this paper. If nothing else, this work could eliminate these as likely causes of the sinking. The third is very difficult to assess now but the possibility of it could be informed somewhat by the structural analysis. Extreme freak waves do occur and have the properties of being very short lived and very local and so only Victory in the fleet might have been hit. The physics of how such waves form is not well understood and so it would be very difficult to create a credible model of such a wave.

4. THE DESIGN STORM

The ship foundered in an extreme storm and, of course, there is no quantified information from the time which describes that storm, or, more importantly, describes the sea-state.

So, rather than try and guess storms, the approach has been to consider what might be the upper-bound of the conditions, based on the physics of what could happen.

The ship sank at a point mid-way across the English Channel south of Start Point. This point is well on to the English/French continental shelf where the water depth is about 70 m and the surrounding sea-bed is fairly level. The storm in the channel would almost certainly be caused by a large depression north of Ireland, causing strong winds over a long fetch out into the Atlantic. The winds in the Channel would have been between west and south west, with associated large waves flowing up the Channel.

This scenario leads to a fundamental wave length at the sinking site of probably about 136m. This wave would be a wave transitioning from a long deep-water wave out in the deep ocean into a shallow-water wave. This transitioning starts to occur once the water depth is half the wave-length: longer waves coming into shallow water tend to shorten in wave length and increase in wave height.

The maximum wave height (crest to trough) possible for stable waves is about 7% of wave length, which gives a wave height of 9.52m. Because of the effect of the depth getting shallower, and the presence of local strong winds, the wave form is probably close to trochoidal, with breaking crests.

A 136m long wave has a speed of 11.93 m/sec, a period of 11.4 sec and a frequency of 0.088 Hz, and a particle velocity at the crest of 2.63 m/sec. The coexisting mean wind speed is taken to be 20.6 m/sec.

5. THE SHIP IN THE STORM

The last sighting of the ship describes it as "wallowing", which implies sinking while upright, rather than something more dramatic like a broach or a capsize. In this case, two scenarios seem to be possible, namely;

1. the ship suffered structural failure of the hull under bending and shear from the large waves, leading to massive leakage; probably the worst load case for this is with the crest of a wave at midships which would add extra hogging to the hogging inherent in such a ship

2. the ship was flooded by waves smashing the stern lowest gallery and pouring water into the un-partitioned interior.
3. gallery was lower than was normal; the dynamic pressure of sea-water arriving with a particle velocity of 2.63 m/sec is about 0.35 tonnes/m2.

In both these scenarios, the ship is taken to be running down-wind; it is difficult to estimate the down-wind speed of the ship by fluid mechanics calculation and it may well have deployed sea-anchors anyway. Figures 1 and 2, below, show these two scenarios.

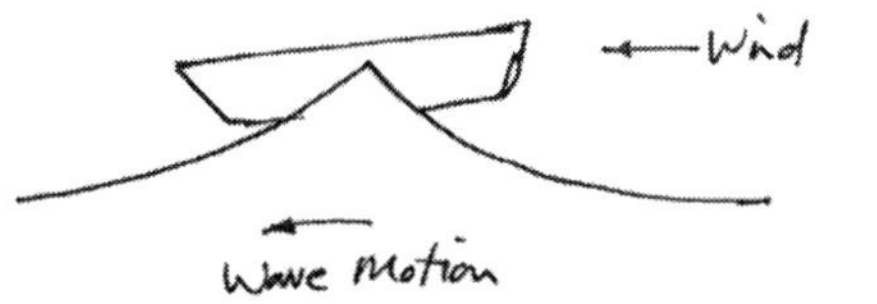

Figure 1: Scenario 1

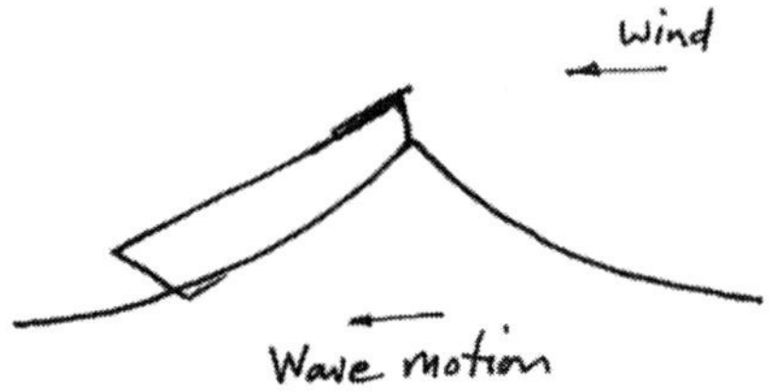

Figure 2: Scenario 2

6. GLOBAL BENDING AND SHEAR MODEL

The first task in developing a computer model of Vitory 1737 is to digitise the ship. Using the lines plans, the software tool DelftShip is used to capture the detailed shape of the ships hull (See figure 3). From this data a CAD model can then be developed (See figure 4).

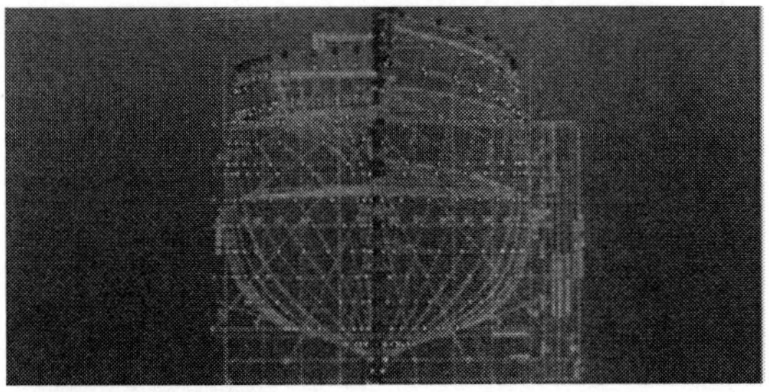

Figure 3: Lines plans

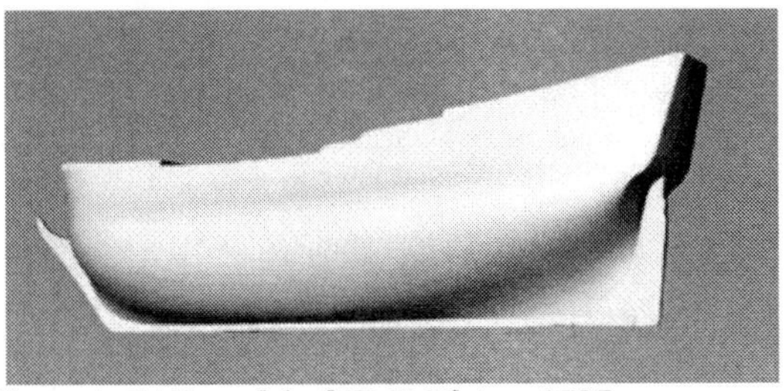

Figure 4: CAD model of HMS Victory 1737

The global loads in the hull are then derived using the Bonjean method in which the hull is represented by a single linear beam and the buoyancy of the hull is modelled as a series of highly non-linear springs. This type of model is shown in figure 5.

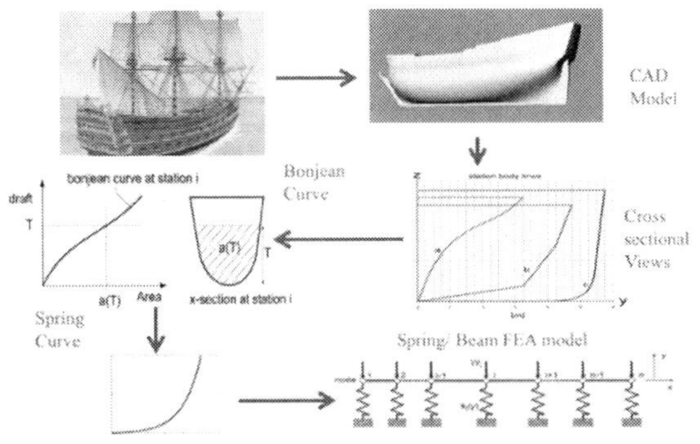

Figure 5: Bonjean Spring/Beam Model

The application of the wave for the two scenarios can now be undertaken where the ship is represented as a beam, and the wave profile by Bojean springs.

This model is a static model in that the dynamic nature of the waves and the ship are not modelled. The justification for this approach is that the pitching frequency of the ship is about 0.16 Hz, which is about twice the wave frequency, and the radiation damping is estimated to be 16% critical damping. Hence, un-damped resonance with waves is not likely.

It is, of course, possible to adapt the Bonjean model for dynamic analysis by applying traversing wave motion to the grounds of the Bonjean springs. The structural properties of the hull beam are calculated assuming that the hull is an integral solid skinned structure rather like a huge dug-out canoe.

7. ANALYSIS OF STRUCTURAL STRENGTH

The ship is constructed of fairly massive pieces of timber fastened together with metal bolts and treenails. Metal bolts in clearance holes are good for purely tensile loads; treenails can carry tensile loads but are really good at carrying shear loads with very little slippage.

For the behaviour of the hull as a shell structure carrying global bending moments and shear, the strength of the treenails is critical and governing. This applies to the relatively simple load-path at a strake joint where tensile bending stress in one strake taken out through a frame and applied to the continuing strake; more importantly, it applies in a vital way in the complex lozenging interaction between strakes (and linings) and frames which allow the hull to carry shear loads and so behave as an integral shell. Without this interaction, the hull would simply be a bundle of sticks.

The bending stresses and shear stresses in the hull shell are calculated from the global bending moments and shear forces using standard engineering bending theory. As an aside, this is the Euler-Bernoulli bending theory which was published in about 1750. Using strength of material calculations, the shear force in any treenail under the particular loading condition can be considered. The onset of structural failure, probably defined as enough structural disruption to allow uncontrollable leakage, is when treenails reach a critical load which is close to the ultimate failure load.

From an engineering perspective, the structural configuration if these tree-nails is inherently poor. Later wooden ships had triangulation to improve the structural behaviour and the ultimate expression of triangulation is in the composite clippers with their iron chassis.

7.1 USEFUL STRENGTH OF TREENAILS

A treenail fastening is a complex structure in its own right; there is bending and shear in the dowel itself and bearing stresses on the surrounding timber. BS 5268 – Structural Use of Timber does not cover this configuration of materials but it does offer 2.0 N/mm2 as a grade stress in shear for oak. Now the grade stress is a measure of stress at which the element could operate normally in, say, a building, i.e. a safe working stress. It is not a measure of the ultimate stress at which the element is starting to fail. It would be reasonable to expect a factor of about 2.0 to 2.5 between grade stress and ultimate stress.

Probably a more useful source of information comes from Creuze ("Treatise on the Theory and Practice of Naval Architecture" Augustin Creuze, 1840) in which he describes his load tests to failure on various treenail fastenings. Taking his typical results, and expressing and summarising them as a single value of shear stress at ultimate failure, gives a figure of about 5.5 N/mm2. This is not inconsistent with BS 5268 grade stress multiplied by 2.5. Creuze does observe that his trial joints were surprisingly ductile for the most part, with joints moving ½" before failure. Such a displacement might be more than enough to allow uncontrollable leakage and so for this analysis it is proposed to take 75% of Creuze's shear stress as the effective onset of failure of a treenail in a floating ship. This is 4 N/mm2.

7.2 INITIAL PRE-LOAD BETWEEN PLANKS

The lozenging model indicates the well-known importance of pre-load between adjacent strakes, whether this is caused by moisture content expansion or caulking, or a combination of both. It is impossible to derive this pre-load by calculation as the range of possible values is huge. So the proposal is to measure it in-situ on suitable wooden ships which are afloat. The method is an adaptation of the Centre-Hole Technique first developed by the CEGB for measuring stress in turbine casings. In its original form, it involves drilling a small blind hole in the centre of a strain gauge rosette and deriving the initial state of stress from the strain changes in the strain gauges. Our adaptation, and simplification, is to cut a small, blind slit and measure the closing of the slit with a Demec gauge. This method needs calibrating with a small FEA model.

8. CONCLUSIONS

Although much has been reported on HMS Victory 1737, the reason for her sinking still remains a mystery. This multi-disciplinary project underway between National Maritime Museum and the University of Greenwich is aiming to address a number of questions that could shine some light as to why the ship sunk, and also how the latest scientific methods of the time could have influenced ship design in the British and French fleets of the time.

9. ACKNOWLEDGEMENTS

The authors would like to acknowledge the resources provided by both the National Maritime Museum and the University of Greenwich to support this project.

10. REFERENCES

1. Rajaguru, R., Mason, P., Bailey, C., Stoyanov, S. 'Buckling analysis on the hull of the historic paddle steamer – Medway Queen' *Ships and Offshore Structures*, 2013.

11. AUTHORS BIOGRAPHY

Catherine Tonry holds a research fellowship at the University of Greenwich. She has recently completed her PhD, and has extensive experience using computer-aided technologies, including previous work presented at RINA detailing models for the Cutty Sark.

Peter Mason is a Visiting Professor at the University of Greenwich. He is an experienced engineer who worked on the Cutty Sark Conservation Project and the Medway Queen Project. He is also a Royal Academy of Engineering Visiting Professor a Surrey University.

Mayur Patel is professor of Computational Fluid Dynamics at the University of Greenwich. He has worked on modelling hydrodynamics of the of historic ships – recently the Cutty Sark.

Wyn Davies is a self-employed naval architect specialising in maritime heritage projects, currently acting as a mentor for the Heritage Lottery Fund. To date he has worked on some 37 historic vessels as technical advisor or HLF Monitor/mentor, given around 15 papers at national and international conferences and has three books to his credit.

Chris Bailey is Professor of Computational Mechanics, and Director of the Computational Mechanics and reliability Group at Greenwich. He has managed a number of modelling projects related to historic ships including Cutty Sark and Medway Queen.

Nick Ball is assistant curator of the ship model collection at the National Maritime Museum. he is a member of the Joint Nautical Archaeology Policy Committee, and is lead curator on the redisplay of the Voyagers Gallery at the NMM.

Simon Stephens is curator of the ship model collection at the National Maritime Museum. He is responsible for maintaining the ships models at NMM, as well as their records.

CAPSIZE OF *SS DAPHNE* ON ITS LAUNCH, 3rd JULY 1883 - A TECHNICAL REAPPRAISAL

I E Winkle, MRINA (Retired), UK

SUMMARY

Sir Edward Reed's Inquiry [2] discounted the actual launch process as contributing to the disaster. This assumption is challenged in the light of testimony given at the Inquiry as well as letters written to Francis Elgar by Alexander Stephen [11], identifying heavy loose items on board together contributing a greater potential effect than the personnel present at the time - Reed's primary cause. This paper demonstrates the capsize resulted from four contributory factors on top of relatively low stability. These include an initial offset centre of gravity, a large varying heeling moment generated by the drags and anchors, the subsequent movement of loose weights **and** the 195 personnel and their effects.

This disaster had a profound effect on the UK shipbuilding industry, introducing systematic stability calculations aided by the contemporary introduction of the Amsler-Laffon Mechanical Integrator and precipitating the foundation of the world's first civil academic chair in Naval Architecture.

1. INTRODUCTION

A general review of Daphne's capsize, including Figures 1 and 2 showing her general arrangement, was presented in a RINA paper by Fred Walker [1] in 2002, but without a detailed technical appraisal. For this we must rely primarily on the Parliamentary Inquiry held in Glasgow for 5 days from 10th July to 31st July before Sir Edward J Reed KCB, FRS, MP, as Commissioner, "to throw light upon the actual cause of the accident". The report [2] of that Inquiry forms the only reliable reference to the facts and observations made at the time despite a huge collection of newspaper cuttings [3] relating to the incident. The report is in four parts:

- Sir Edward Reed's general review of the case – Part I
- His technical treatment of the issues relating to ship stability in general and *Daphne* in particular – Part II, in which are found the only estimated hydrostatic data (Figure 3) and stability curves (Figure 4) relating to *Daphne*
- A record of the proceeding of the Inquiry, in which the individual testimony of witnesses is recorded. Reference is made to a large number of technical drawings and other exhibits which were presented to the Commissioner but none of these appear to have been published or archived.
- An Appendix of communications with the Clyde Navigation Trust on the subject of bye laws and procedures relating to the safety of ships during a launch – in particular the need to suspend passing traffic for the duration.

2. INPUT OF 4 NAVAL ARCHITECTS

Sir Edward Reed was himself an eminent Naval Architect having risen to the post of Chief Naval Constructor in 1863 at the age of 33 resigning in 1870 in a bitter dispute opposing the building of *HMS Captain* which foundered a year later with the loss of nearly 500 lives. He published a number of books, one of which 'Treatise on the Stability of Ships' [4] was published in 1884 shortly after his contribution as Commissioner investigating the capsize of *Daphne*.

Apart from representatives of the owners (Glasgow & Londonderry Steam Packet Company) and employees of Alexander Stephens shipyard who were responsible for the build and launch of *Daphne*, three eminent Naval Architects of the day appeared to give evidence relating to launch procedures and stability considerations. Between 3rd and 11th June, the first, Francis Elgar, Fellow of the Royal School of Naval Architecture & Marine Engineering and Member of Council of the Institution of Naval Architects (INA), was employed as a consultant by Alexander Stephen to assess the hydrostatics and stability of the vessel as launched. He initially estimated the post-launch draughts as 5ft. 1.5in. at the bow and 8ft. 10.5in. at the stern giving a mean draught of 7ft. (at amidships) corresponding to a displacement of 401ton [1 p38]. Given that all original data for *Daphne* was in **Imperial Units**, these are used throughout the analysis in this paper. Thus, the following hydrostatic data were estimated for this condition (all above top of keel, K, 0.625ft. above base):

KB = 3.67ft.
BM = 7.75ft., giving
KM = 11.42ft.
KG = 10.76ft. estimated, giving
GM = 0.66ft. (nearly 8in.).

This condition took into account an estimated 200 men and boys and their tools (at 16.67 per ton) weighing about 12ton in all, distributed 8ton at 2.7ft. above the bridge deck and the remainder acting just under the main deck (some being on that deck and the remainder in the hold). He estimated the effect of the missing boiler (with centroid given as 8.5ft. above top of keel) on the KG of the vessel to be more than offset by the funnel and coamings still to be erected.

No data is given as to the size or exact position of the missing boiler in the Inquiry Report. However, the Specification [5] for *Daphne* alludes to a single, 90psi,

Scotch Boiler with 3 furnaces 14ft. OD and 11.5ft. long. No weights are given but it required a larger opening in the deck and casing. Why this should be offset to the starboard side is not explained although may relate to the outreach of the crane to be used at the Finnieston Quay where it was to be shipped.

The next to give evidence [2, p42], John Harvard Biles, was Naval Architect to Messrs J & G Thomson (later John Brown), Clydebank who was questioned on the capsize at launch (only about 8 months earlier) of the *Hammonia*, a passenger vessel built by J & G Thomson for the Hamburg America Line, at 3969 ton gross, 373ft. long and 44.9ft. beam (about 8 times larger than *Daphne*). As no-one was killed in this incident there was little publicity and no Inquiry, but Biles described a sequence of events that was remarkably similar to the *Daphne* ending up with the vessel listing at 40-50 degrees to starboard. The ship was towed to dry dock, righted and a careful inclining experiment performed to determine its GM and KG from which its 'as launched' Stability Curve was determined. This was offered for comparison with *Daphne* having a GM of 0.6ft. (close to Elgar's estimate for *Daphne*). This curve is shown as Fig. 12 [2] in the Commissioner's final Technical Conclusions together with Fig. 13 (reproduced here as Figure 4), his recalculation of *Daphne*'s – both being remarkably similar in character and scale. The fact that both vessels were launched at similar states of the tide, *Hammonia* from the north bank and *Daphne* from the south, each capsizing toward the direction of tidal flow appears to have been highly significant to the both the Commissioner and Biles, from which much discussion of the impact of river currents on launching proceeded, to little positive effect.

Benjamin Normand, an eminent Naval Architect from the Le Havre shipyard in France, went into a considerable discussion [2, p50] of the transition from the point at which the fore poppet drops off the end of the standing ways as a point of likely short-term instability due to the large vertical load acting below the keel which is rapidly replaced by the stable situation once the vessel has dropped to its correct draught and trim a few seconds later. In some cases vessels can be observed to oscillate at this transition point depending on the speed and nature of this release. If the vessel twists slightly on the ways, one side of the fore poppet may drop before the other and a roll may result. He also goes on to point out that a vessel built out in the open for some time will undoubtedly accumulate a significant quantity of rain water which may well have an influence on the stability through what is now referred to as 'free surface effect'. This was not picked up by anyone, professional or otherwise, at the Inquiry.

After a period of 20 days, Francis Elgar was recalled [2, p56] to present his calculated launch stability curves and hydrostatic curves [2, Fig16] - see Figure 3 - based on information additional to that on which his earliest estimate was made. His original mean draught estimate was revised to 7ft. 5in. and the GM to 6.5in. Figure 3 indicates the effect of the varying assumptions relating to draught and displacement. The three superimposed stability curves of Figure 4 consider the Righting Arm, GZ, at angles up to 65 degrees for the conditions that:

- there were neither effective bulwarks nor a watertight poop, and
- each of the conditions in which the poop and bulwark could be considered fully watertight.

Most discussion suggests that the bulwarks (although to the height of the Bridge Deck) had significant leakage potential through their port side scuppers (which may not have their externally opening, hinged, closing plates properly secured) onto the main deck. At this point the number of persons on board was noted finally as 195 of whom 67 were below the main deck. In the event, the Commissioner later modified Elgar's stability curves which he presented in his technical summary [2, Fig13] – see Figure 4 - for a reduced GM of just under 4in. - following an inclining experiment conducted by him after the vessel was recovered, but which may have contained significant quantities of river sediment as well as 'saturated' wood decks.

In his final technical conclusion [2, p19] Reed recommended the use of the recently introduced Amsler Integrator as a quick and reliable method of calculating the GZ Stability Curves for all ship loading conditions as described in an 1880 [6] by C W Merrifield and in 1882 [7] by J H Biles. This instrument had been developed theoretically by Mr J Amsler-Laffon in Switzerland shortly after the planimeter (1855), first constructed for the International Exhibition in Paris in 1867 but not introduced into the UK until 1878 [8], being quickly adopted for this purpose by the Denny shipyard in Dumbarton amongst others [9]

3. COMPARISON WITH IMO STANDARDS

Once shipbuilders had the Integrator, it soon became standard practice to construct Cross Curves of Stability for a vessel from which a GZ-Θ stability curve can be generated for any loading condition for which the displacement and KG are known. By then, it was also well understood that as well as representing the righting lever, GZ, at any angle, this curve also indicates the energy required to heel the vessel between any angles, by multiplying the enclosed area of the curve between those angles by the vessel's displacement. Such curves now form the basis for IMO Stability Regulations which stipulate the *minimum* operational requirements for such a curve of stability in terms of:

- GM (gradient of the curve at origin when plotted at 1 radian = 57.3 degrees) $\geq$ 6in.
- Maximum GZ and angle at which it should occur $\geq$ 8in. at $\geq$ 30 degrees
- Area under the curve from:
  0-30 degrees $\geq$ 0.180ft.radians

0-40 degrees $\geq$ 0.295ft.radians
30-40 degrees $\geq$ 0.098ft.radians.

As Reed gives the curves for the launching conditions of both *Hammonia* and *Daphne* as Figs 12 & 13 (Figure 4) in [2] it is possible to draw comparisons in Table 1 with the above standard which is now regarded as the *minimum* for the safe operation of any commercial vessel in *sea-going* conditions.

It can be seen that, given Reed's assumptions (which are not spelled out, other than the Displacement had now increased to 521ton with a GM just under 4in.), *Daphne* was significantly deficient in stability in the range up to 30 degrees and had only about two thirds of the GM and GZ_{max} values now considered the minimum for sea-going conditions. However, to heel to 51 degrees, at which point she would flood catastrophically, would normally require a heeling moment of about 134ton.ft. assuming that her actual stability curve ranged midway between curves B and C in Figure 4 – allowing for some progressive flooding through the scuppers in the bulwarks. She was undoubtedly a tender vessel in this condition and would experience a steady angle of heel of about 10 degrees from the upright condition under the action of a moment of only 32ton.ft. to port. At 12.2 men/ton transferred though about 11.5ft., this would require only about 33 more men (and their equipment) to be near the port side than near the starboard side (i.e. 81 to starboard and 114 to port). If this was the case then she would easily 'lurch' to about 16 degrees dynamically from the upright before attempting to right herself, as suggested by many of the witnesses. This seemed an altogether plausible assumption given the fact that portions of the starboard decks were missing. This heeling moment could also be caused by the distribution of loose material which would produce roughly the same effect if the vessel was suddenly released from the upright. However, this initial 'lurch' must also have been sufficient to send a large proportion of the personnel on board involuntarily tumbling toward the port rail which would produce the necessary moment transfer to take the vessel dynamically towards 30 degrees, at which point all the equipment and loose materials on board would have also slipped to port to complete the capsize. An important assumption therefore is that there must have been a short term dynamic initiator (holding the vessel near upright) which when released acted together with an offset centre of gravity to roll the vessel initially to at least 16 degrees which then set off a chain reaction of further substantial heeling moments to port. None of this is suggested directly in Reed's discussion and there was no attempt to represent the Stability Curve in this condition.

4. ANALYSING THE LAUNCH PROCESS

The general procedure and arrangement of drag chains and anchors to stop the vessel are described in some detail by John Stephen (partner and brother of Alexander Stephen), the only partner present at the launch, who was responsible for the practical work of the shipyard. He appears to have witnessed the launch from the right hand side of the launch platform – the port side of the vessel. He was well positioned to see the sequence of events leading up to the final capsize, which he described [2, p23] thus: "On the vessel leaving the ways she listed slightly over to port, but recovered herself completely when brought up - then inclined to port again – again recovering to upright, and a little over to starboard, and then hesitatingly went over entirely to port. He gave a detailed description [2, p22-23] of the drag chains and anchors used to arrest *Daphne* once in the water and these have allowed a 5 stage scenario of the motion of the vessel once it cleared the ways to be considered.

Drag chains were only attached to the starboard side of the vessel by a chain cable at a point 71ft. 6in. aft of the stem with the shackle 12ft. 10in. from the centre line on the side shell in way of the main deck about 13ft. 3.5in. above keel (identified from a photograph of the capsized vessel) – shown in Figure 2. The drag chains were about 17ton and would therefore exert a frictional drag of about 10ton (assuming a coefficient of friction of about 0.58). They was secured at their landward end to a 12cwt. anchor with one fluke buried in the ground which was backed up by a further drag of (2ton 6cwt.) plates attached to the upper fluke (to keep the anchor buried). The port side had no drag chains, but used a chain cable connected through the port hawse pipe (near the centreline at the stem and just above the main deck) to a 7cwt. anchor backed by 2ton 5cwt. of drag plates attached to it upper fluke. It has been difficult to estimate the frictional drag of the two anchors and their drags, but for the purposes of this discussion they are assumed to provide an additional 6ton to starboard and 4ton to port. The starboard drags activated about 37ft. after the vessel dropped off the ways (at which point the bow would have been about 13.5ft. clear of the end of the ways and continued for 90ft. thereafter, the last 15ft. 7in. of which included the starboard anchor. The port anchor dragged for 56ft. 3in. which implies that it started to act 33ft. 9in. after the starboard drags. Figure 5A illustrates the drag loads in the first 4 Stages. The 5 stage scenario alluded to above, now takes the form of:

- **Stage 1**: 0-37ft. after leaving ways – no drags or anchors, only influence of tidal current
- **Stage 2**: the next 33.75ft. of motion under the influence of 10ton of starboard drag force and the tidal current
- **Stage 3**: the next 40.67ft. of motion under the influence of 10ton of starboard drag force, 4ton of port anchor force and the tidal current
- **Stage 4**: the final 15.58ft. of motion under the influence of 16ton of starboard drag and anchor force, 4ton of port anchor force and the tidal current
- **Stage 5**: stationary, but drifting with the tide while moored to the shore by the port and starboard anchor and drag systems. In this state

John Stephen stated that the bow ended 30ft. and the stern 85ft. upriver of the centreline of the launch ways.

An additional observation given by Robert McMaster – the Linthouse yard Manager, who was standing right under the bows of the ship on the starboard side - was "when once she was half down the ways, her stern getting out well into the river, the flood tide commenced to slew her stern up the river. I was in a position where I could see that, and I noticed it specially. My attention was drawn to it." This resulted in the ribbons on the standing ways "rubbing hard".

An estimate of speeds during each phase has been derived from an average of the speeds in the last 150ft. of launching contained in a graph (Figure 6) by J H Biles for a number of contemporary launches at the J & G Thomson Yard in the discussion to a paper by W. Denny 'On Launching Velocities' [10].

Stage 1 therefore commenced with the stern swinging somewhat to port (up river) so that the starboard sliding way is likely to have dropped off the standing way fractionally earlier than that on the port side. This would have given a very slight kick to roll the vessel to starboard before it commenced its observed heel to port. This stage would have taken about 3.2 seconds (assuming an average speed of about 11.5ft./sec.) before entering Stage 2.

Stage 2 sees the take up of the starboard drag acting 12ft. 10in. off *Daphne's* centreline so that a turning moment was generated tending to rotate the stern down river. This stage would take about 4.0 seconds if we assume an average speed of about 8.5ft./sec. The turning moment would continue, gradually reducing to zero when an equilibrium position was reached as the line of action of the starboard drag passes through the intersection of the centreline and the centre of lateral resistance (CLR) which is estimated to be 25ft. aft of the starboard drag shackle and about 2.3ft. above keel (lowered by the sliding ways underneath) with the vessel trimmed by 3.75ft. by the stern, see Figure 2. At equilibrium, the stern would swing to starboard so that the drags were acting some 27.2 degrees away from the centreline giving a *maximum* component at right angles to the centreline of some 4.57ton. This would be resisted by an equal and opposite force acting through the CLR to give a vertical lever of 11ft. between the drag force and CLR. This would generate a *maximum* starboard heeling moment of 50.3ton.ft. To this should be added the moment effect of the drag cable itself. If you assume the unsupported length to weigh about 1ton, half of this will act vertically on the ship 12.83ft. to starboard of the centreline adding an additional 6.4ton.ft., to give a *maximum* starboard heeling moment of about **57ton.ft.**. It is not clear whether 4 seconds would be enough time to generate this maximum moment before the port anchor line applied its drag force to the bow at the beginning of **Stage 3**. Figure 7 illustrates the components of the drag forces acting to heel *Daphne* in this and the two following stages.

Stage 3, under the action of two roughly parallel drag lines, would impose a 14 ton braking force intersecting the centreline at the CLR (at equilibrium) which would reduce the angular displacement of the net drag force to some 11.4 degrees from the ship's centreline. The port anchor force through the lowest point of the hawse pipe is assumed to be about 4.35ft. above the starboard drag connection, taking trim and sheer into account, acting about 1.5ft. off the centreline to port and about 4ft. aft of the stem. The resulting starboard heeling moment would now reduce to [1.98x11+0.79x15.35] = 33.8ton.ft. Adding in the effects of about 1ton of drag cable to both port and starboard adds about 5.65ton.ft. to starboard to give a total equilibrium moment of about **40ton.ft.** to starboard, somewhat less than the maximum under Stage 2. This stage lasted for some 8.1 seconds assuming an average speed of about 5ft./sec.

Stage 4, which saw the starboard drag force increase to about 16ton, lasted for about 12.5 seconds, assuming an average speed of about 1.25ft./sec. In this condition the angular offset of the drag cables to the centreline at equilibrium would become about 14.6 degrees and the resulting heeling moment to starboard would increase to about **65ton.ft.,** including the moment effect of the drag cables. Figure 5B illustrates the progressive heeling moment caused by the drags and anchors, plotted against both distance and estimated time. The overall time from leaving the launch ways to floating stationary in the river is less than 30 seconds based on these estimates.

Once the drags had stopped her, *Daphne* would rebound slightly as she entered **Stage 5**, the starboard heeling moments, estimated above, would have been almost instantaneously removed as she started to drift up river in the current under the mooring action of the drags only. The sudden removal of a substantial heeling moment to starboard would initiate an immediate roll to port to a point at which the initial roll energy had been absorbed by rolling nearly as far to port as she had originally been rolled to starboard from her equilibrium position. She would then have commenced a rebound towards starboard so long as nothing shifted on board.

Much of the observer evidence supports the view that she was close to or just beyond the vertical to starboard at the end of Stage 4. It is clear therefore that, at that time, her centre of gravity had a significant offset to port. Assuming that the 65ton.ft. moment of Stage 4 brought her close to upright, the centre of gravity seems to have been about 65/521 = 0.125ft. (1.5in.) to port of the centreline (assuming Reed's final displacement of 521ton). This is equivalent to an initial offset of about 61 personnel and their equipment between port and starboard (assuming all other material was evenly distributed about the centreline) – 90% more than Reed's suggestion in Section 4 above. It is now possible to reconstruct the actual stability curve for this condition by raising the baseline of Figure 4, by an amount 0.125cosΘ, this being the reduction in Righting Lever, GZ, to port for this condition. Examining this modified curve, Figure 8, will allow us to understand what happened during Stage 5. In particular it can be seen that the crossing point of the curve and modified baseline is now at about 15.6 degrees to port which becomes the *initial* equilibrium heel angle. It may also be useful to try to estimate the roll period corresponding to this modified GZ curve, with a GM of about 4 inches, to gain some insight into the time taken to execute each rolling action.

An estimate of the roll period is given by:

$$\mathbf{T_0} = C.B/\sqrt{GM} = 0.47x25/\sqrt{0.333} = \mathbf{20.4sec.}$$

Based on this estimate, it would seem that in 3.2 seconds, the initial roll to port in **Stage 1** might have been about 8 degrees, reaching a maximum of perhaps 16 degrees as the drag action in **Stage 2** checked it, thereby limiting the potential for major shift of material or personnel during this short initial phase.

5. THE FINAL CAPSIZE

Considering the redrawn Stability Curve in Figure 8, in which the large drag moment to starboard, in **Stage 4**, holds the vessel upright. As soon as that moment is released, it would take about 10 seconds to roll from that position past the equilibrium angle (15.6 degrees) to achieve an energy balance at the point where the vessel was seen to check its motion before attempting to roll back again towards the upright. This check angle would be about 28 degrees assuming a small amount of roll damping. This is deduced from the fact that the area A1 is proportional to the potential energy stored in resisting the drag forces which, once released, must be recovered by rolling the vessel beyond its equilibrium angle though an area A2 which will be slightly less than A1 (due to loss of energy through damping effects). However, by about 25 degrees the personnel, their equipment and any other loose materials would have started shifting to port, far enough to initiate the final roll from about 20 degrees to the point of capsize beyond 51 degrees. The ***minimum total*** mass moment shift to achieve this would be at least 313ton.ft., which would move *Daphne's* centre of gravity to about 0.6ft. to port of the centreline. This would have required a total of at least 27ton of men and materials to have moved an average of about 11.5 ft. It is important to note that in this condition the base line of the Stability Curve has now shifted up (by the lost GZ = 0.6cosΘ) and is now above the sustainable stability curve B (including the watertight poop). Assuming the bulwarks were *substantially* intact against flooding, the stability curve would run just along Curve C in Figure 4 (allowing for some leakage through the 3 port scuppers in the final roll). In this condition the equilibrium heel angle has reached about 36 degrees, which is itself an unsustainable condition, as progressive flooding would undoubtedly occur by this point. Finding herself at about 20 degrees with a new, larger, positive potential energy (proportional to area A3), *Daphne* would now roll rapidly to port with a shorter period of about 10 seconds (having a much increased local GM) as it attempted to reabsorb the potential energy of A3 in area A4. Capsize is guaranteed if the area A4 up to 51 degrees is only a little smaller (due to damping) than A3 or if even more material shifts to port at these larger angles. In the event, *Daphne's* last dynamic roll, from about 20 to beyond 51 degrees, ensured water flooded over the bulwark just forward of the poop, from which point there was no residual stability and capsize was inevitable. The ***minimum total*** moment transfer of about **313ton.ft.** to meet these conditions is greater than the *maximum* of 184ton.ft. available with all 16 ton of personnel and equipment stacked close to the port side, but well within the capacity of men, equipment and loose plates and other items still to be fitted. In all probability, the final total heeling moment to port could have been as much as **400ton.ft.**. With such a low righting lever GZ (in the 0-30 degree range) it did not require a particularly large initiating offset moment (through asymmetric structural weight and/or personnel) to roll the vessel dynamically to about 28 degrees, even though the equilibrium position of the vessel at 15.6 degrees would have been quite survivable without this dynamic input.

The estimated heel angles deduced from the drag heeling moments are plotted against time in Figure 9 which illustrates the initial port roll checked by the drag moments followed by the near upright phase until the drags slackened, and then the rapid capsize to port with a possible check around 20-28 degrees – the point at which most loose material would be in motion to port to complete the capsize. Three possible curves at this check point are shown, depending on the angle and rate at which the loose materials shifted.

Some of the loose items on board are identified in letters [11] between the 14th and 30th July from Alexander Stephen to his consultant Francis Elgar. These include:

- 2 x 12cwt. bow anchors on the focsle deck
- 10.25cwt. bow anchor, 4cwt. stern anchor and a 2cwt. kedge anchor on the ceiling in the main hold
- 145 fathoms of $1^3/_{16}$ inch anchor cable (about 5ton) on the focsle deck
- $60^3/_4$ fathoms of $^{13}/_{16}$ inch anchor cable (about 1.1ton) on the main deck
- 34cwt. steam steering gear placed in the wheelhouse on the bridge deck
- the galley house and side coaming of the boiler hatch
- adjacent timber decking to starboard @ $50lb./ft.^3$
- 6 boat beams at $13^3/_4$cwt. and 2 boats and davits at 5ton.

Another factor potentially critical to the rapid heeling phase is the presence of any significant quantity of rainwater trapped on board (as pointed out by Benjamin Normand [2]). The presence of keelsons would have mitigated this effect in the early stages, but the fact that they all had drain holes would allow such water to build up on the port side in the holds and engine room during any prolonged roll to port. For every ton (36ft.^{3}) of water present, there would be an additional heeling moment rising from about 5 - 10ton.ft. as the water migrated to port.

Had the drag/anchor systems been reversed, port for starboard, *Daphne* would have capsized within seconds of the drag chains taking effect due to a substantial port drag heeling moment on top of the initial roll to port. In total *Daphne* probably spent about 48 seconds afloat before capsize – a time which seems to fit within the general testimony at the Inquiry.

6. CONCLUSIONS

The risk associated with carrying large numbers of personnel on board a ship during a launch was immediately mitigated by ensuring that future ships would be launched with only those personnel essential to the launching, towage and mooring of the vessel.

Given the circumstances of her launch *Daphne* was clearly deficient in stability, being unable to sustain the large, potentially mobile, weight of personnel, equipment and loose materials through the **dynamic** stages of the launch cycle. Without the loose weights and personnel, she would undoubtedly have survived her launch, only to be found deficient in stability as she entered service. Having been built as a geosym (on the specification of her future owners) of a larger successful vessel *Brier* (D & W Hendersons, Ltd, No 248, 1882), *Daphne's* stability was assumed to be proportional to that of her bigger sister. However, metacentric stability, as represented by BM_t, is proportional to B^2/T (for Beam, *B* and Draught, T). So any reduction in B gives a potentially much more significant reduction in BM_t. In this case her Gross tonnage was reduced from 710 to 449, her Length from 209.8 to 177ft. and her Breadth from 30.2 to 25.3ft. (a reduction to about 84% in each dimension) and Depth from 15.3 to 13.4ft. (about 88%) so her BM_t would only be about 80% that of 'Briar' assuming proportionate draught changes. Until Francis Elgar appeared to support Stephens at the Inquiry, no stability calculations seem to have been performed on *Daphne* on the assumption that what had worked before with *Briar* would work again. However, their basic geometric error concerning BM_t was compounded by the owners' instance on the same weight of major equipment and outfit components being incorporated in *Daphne* as had been in *Brier*, so increasing her relative top weight and forcing her centre of gravity higher, while reducing her capacity to carry cargo.

7. POSTSCIPT

One tangible result was that the next vessel of similar size to be built for the Glasgow & Londonderry Steam Packet Company, *SS Gardenia* (D & W Henderson Ltd, No 318, 1885), being 461ton gross, 172ft. long and 27.1ft. beam compared to *Daphne*'s 449ton gross, 177ft. length and 25.3ft. beam, gained an extra 1.8ft. on her beam, increasing BM_t by almost 15% - an addition of nearly a foot to her metacentric height (GM) - a very significant increase in stability. Another similarity was that the *Gardenia* was also boilered by Alexander Stephen to suit an identical size of engine (rated 140IHP. rather than 120IHP. for *Daphne*), also running at 90psi. *Gardenia's* boiler was built by Alexander Stephen in 1883 according to Lloyd's Register, around the same time as *Daphne*.

Daphne, renamed *Rose* after her capsize, was completed in September 1883, and taken out of service after a grounding off Millport on 1st Mar 1884, to be re-engined and boilered (with smaller 60IHP. machinery) to start her third incarnation as *Ianthe* in the Eastern Mediterranean under the ownership of John Bell & Co. Had her infamous boiler found a new home? The reduction in power (and the resulting reduction in machinery and boiler weights) and early disposal by her original owners, was probably due to her limited cargo capacity due to poor stability and consequent need to carry substantial permanent ballast (50ton of pig iron plus cement ballast recorded in a letter accompanying the final bill on 11th Sept 1883 [11])! She survived in the Mediterranean under Greek owners from 1890 as *Eleni* until sunk by a mine on 10th Dec 1918.

Daphne capsized shortly after the introduction into the UK of the Amsler Integrator (1878), encouraging its early application to the rapid and accurate calculation of ship stability curves, there was now no excuse not to analyse the stability of any vessel prior to, and during, construction. This is brought home in the INA Transactions as no fewer than 8 of the 21 papers read in April 1884 (including [8 & 9]) were on the subject of Ship Stability. This accident therefore represents a turning point in the application of scientific principles to the design and construction of *commercial* ships in the UK, something that had been underway, on and off, in naval circles for 72 years. Indeed, the three main players in the *Daphne* Inquiry were all academically trained naval constructors. Later in the same year, Isabella Elder (the widow of John Elder, the late owner of what became Fairfield's Shipyard) endowed the first ever university chair at Glasgow University dedicated to the education and training of commercial Naval Architects. It was Francis Elgar (1883-1886) and John H Biles (1891-1921) who would become the first and third John Elder Professors. Their graduates would go into many of the Clydeside shipyards as well as round the world to apply a systematic, scientific approach to the design of ships.

Together with Sir Edward Reed, John H Biles established one of the first Naval Architecture Consultancies in the UK which continued as Sir J H Biles & Co Ltd from the death of Sir Edward in 1906.

8. ACKNOWLEDGEMENTS

Early in 2013, in his position as Volunteer Guide and Researcher at the Glasgow Riverside Museum, the author was approached by the Ship Curator to provide a presentation for a public meeting to commemorate the 130th Anniversary of the event in the Museum which was unfortunately cancelled for lack of funds. However, the material has since been presented three times within the Museum to staff and visitors to complement their display commemorating the tragedy. Much detailed material has been gleaned from Alexander Stephen's letters between him and his consultant, Francis Elgar, held by the University of Glasgow Archive Service [11].

9. REFERENCES

1. WALKER, F. M., 'The Capsize of the *Daphne* in 1883 – The world's worst launching tragedy and its consequences', Learning from Marine Incidents II, RINA London, Mar 2002

2. REED, E. J. & LAMONT, R. P., 'The *Daphne* Inquiry, Note of Proceeding at Inquiry into the 'Sinking of S.S. "Daphne" at Glasgow'; Eyre & Spottiswoode, London 1883

3. University of Glasgow Archives, UGD4/22/1/1-3, Alexander Stephen & Sons Ltd., 'Press-cutting books concerning the *Daphne* disaster'

4. REED, E. J., 'Treatise on the Stability of Ships'; C. Griffin, 1884

5. University of Glasgow Archives, UCS3/10/153, Alexander Stephen & Sons Ltd., 'Specification Files 1862-1966, 279 Rose (*Daphne*)', 1882

6. MERRIFIELD, C. W., 'On J. Amsler-Laffon's Mechanical Integrator'; Trans INA 1880

7. BILES, J. H., 'Curves of Stability of some Mail Steamers'; Trans INA 1882

8. AMSLER, A., 'On the uses of J. Amsler-Laffon's Integrator'; Trans INA 1884

9. DENNY, W., 'On Cross Curves of Stability, their Uses, and a Method of Constructing them Obviating the Necessity for the Usual Correction for the Differences of the Wedges of Immersion and Emersion'; Trans INA 1884

10. DENNY, W., 'On Launching Velocities'; Trans INA 1882

11. University of Glasgow Archives, UGD4/4/3, Alexander Stephen & Sons Ltd., 'Letterbooks of A Stephen 1879-1884'

10. AUTHOR'S BIOGRAPHY

Ian Winkle retired from his post of Senior Lecturer at the Universities of Glasgow and Strathclyde in 2005 having been involved in naval architecture, ship design and ship production subjects since 1979. He has maintained a special interest in ship losses, particularly those involving capsize, since the *Herald of Free Enterprise* tragedy in 1987. For the last four years he has been a Volunteer Guide and researcher at the Glasgow Riverside Museum where he was encouraged to pursue historical research. He now lives in Yeovil.

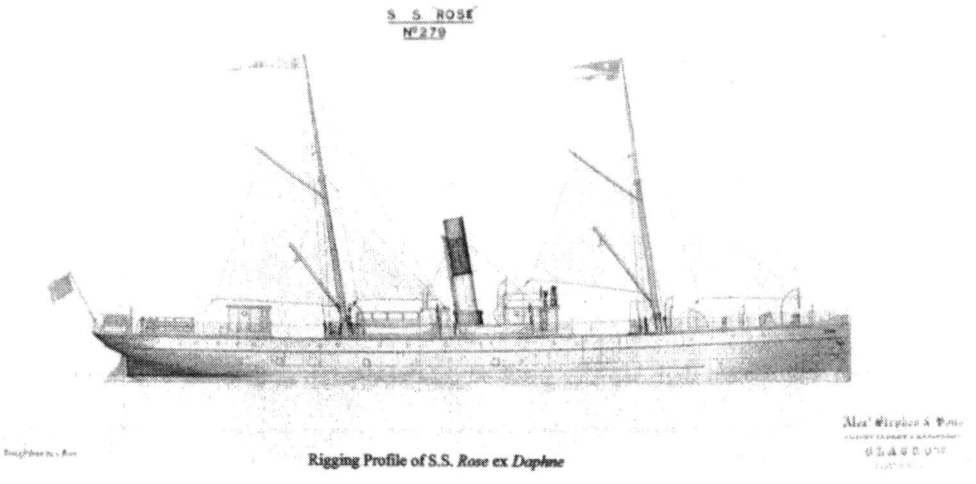

Figure 1: Profile of *Daphne* after completion as *Rose* showing high bulwark between poop and foscle [1]

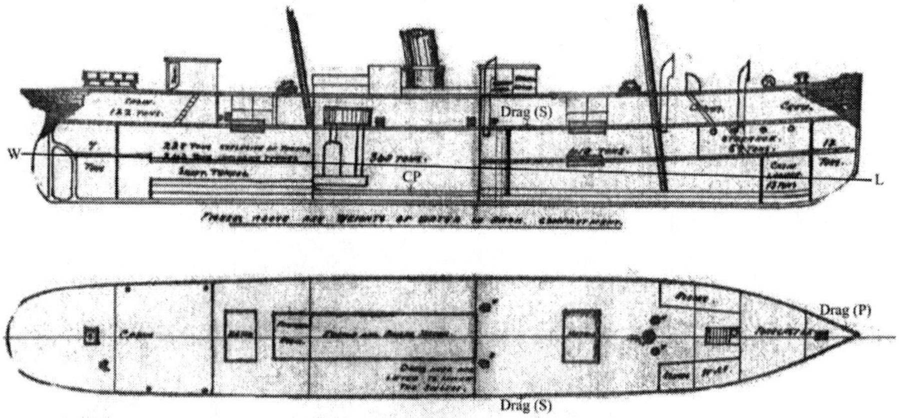

Figure 2: *Daphne* Profile and Deck showing drag attachment points and Centre of Pressure [1]

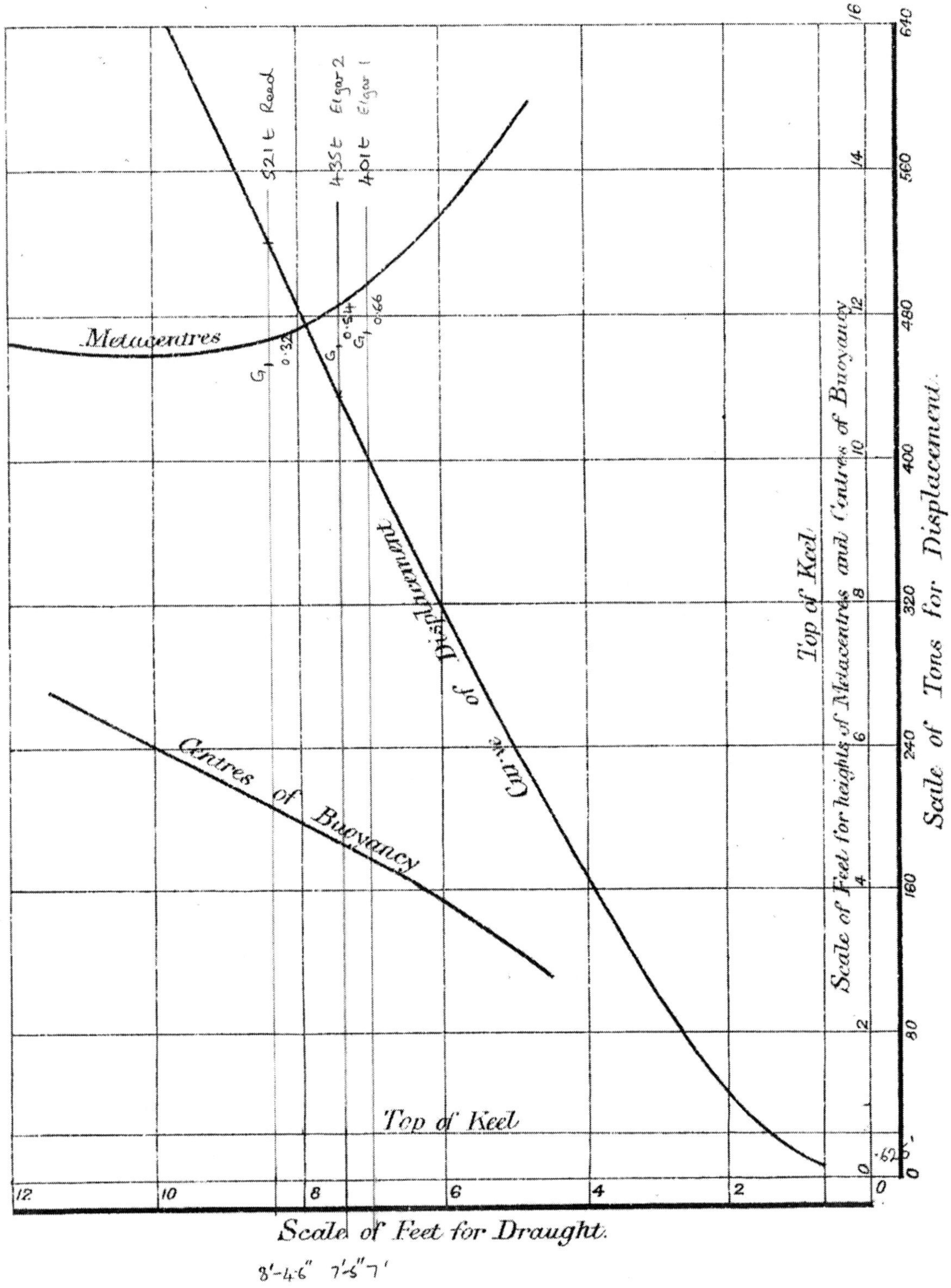

Figure 3: *Daphne* Hydrostatics Highlighting the Three Conditions Referenced in the Inquiry [2, Fig 16]

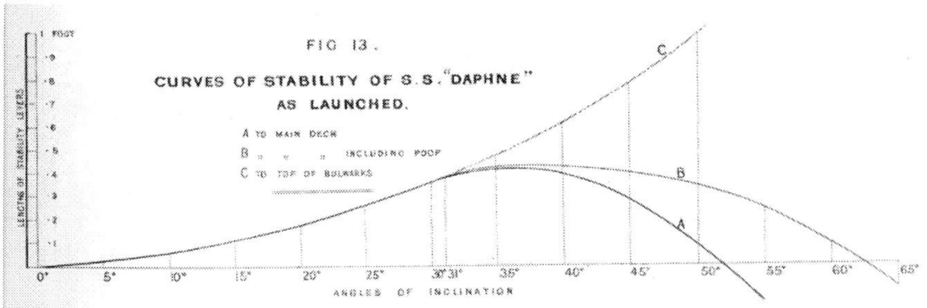

Figure 4: Reed's GZ curves for Daphne after recovery at Displacement of 521 tons [2, Fig 13]

Table 1: Comparison of stability of *Daphne* and *Hammonia* with IMO standards

	Daphne		*Hammonia*	
	Fig 13 [2]	**% IMO**	**Fig12 [2]**	**% IMO**
GM (in)	3.89	**65**	7.47	125
GZmax (in)	5.2	**65**	4.62	**58**
Angle GZ max (degrees)	37	123	32	107
Area 0-30 (ft radians)	.0688	**38**	.0988	**55**
Area 0-40 (ft radians)	.1509	**51**	.1405	**48**
Area 30-40 (ft radians)	.0821	**84**	.0407	**42**

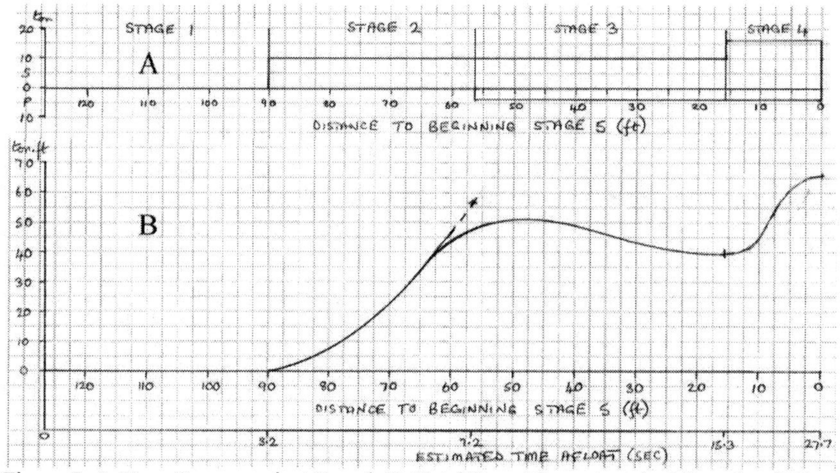

Figure 5: **A**, Drag Forces against Travel; **B**, Total Drag Moment against Travel and Time

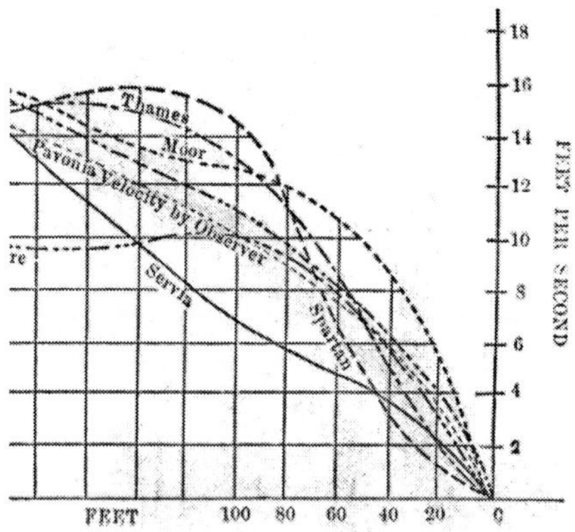

Figure 6: Extract from Figure of Launch Speeds at the J & G Thompson Shipyard givenby J. H. Biles in his Discussion to [10]

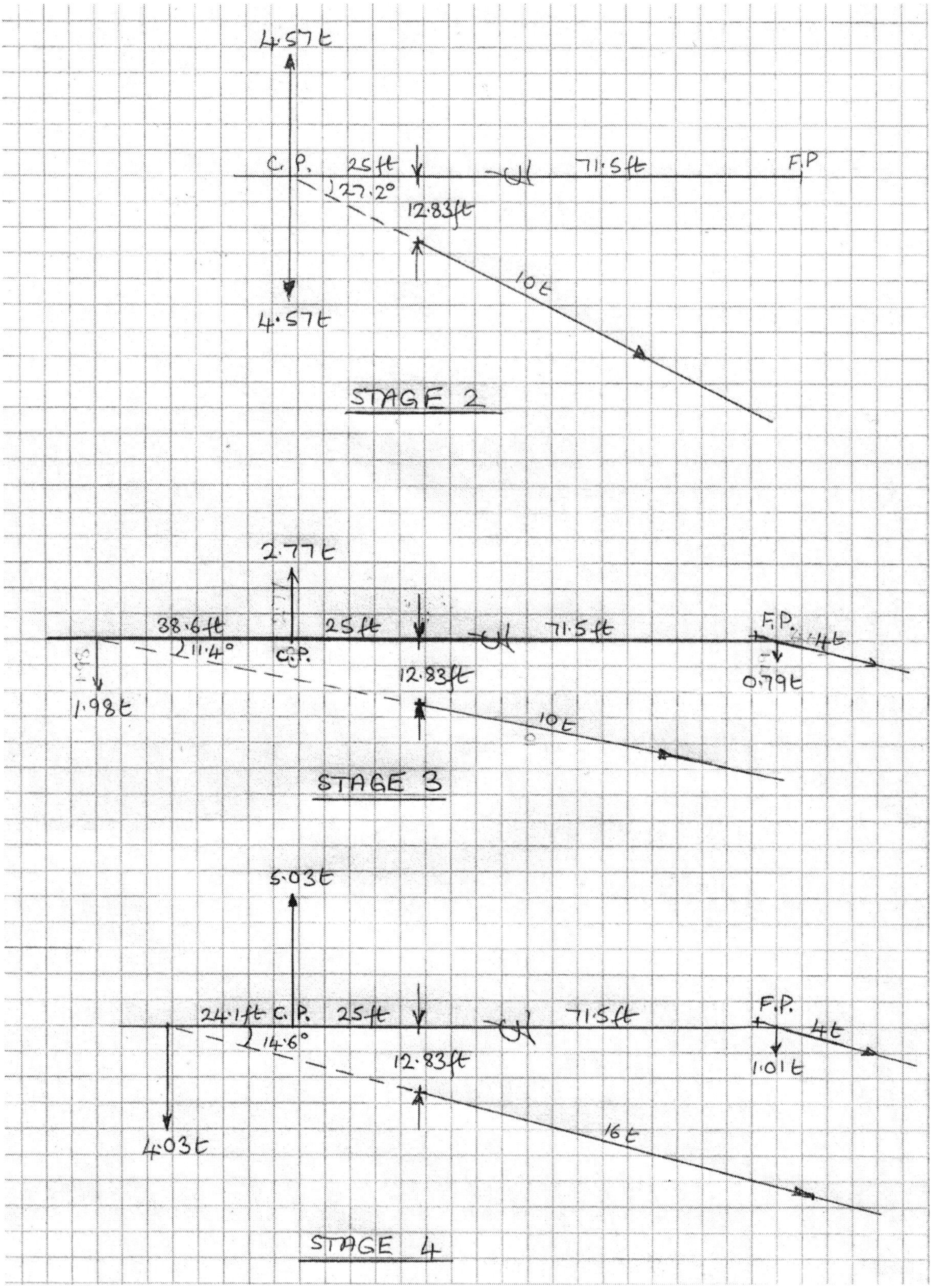

Figure 7: Resolution of Drag and Anchor Forces at Equilibrium for each Stage

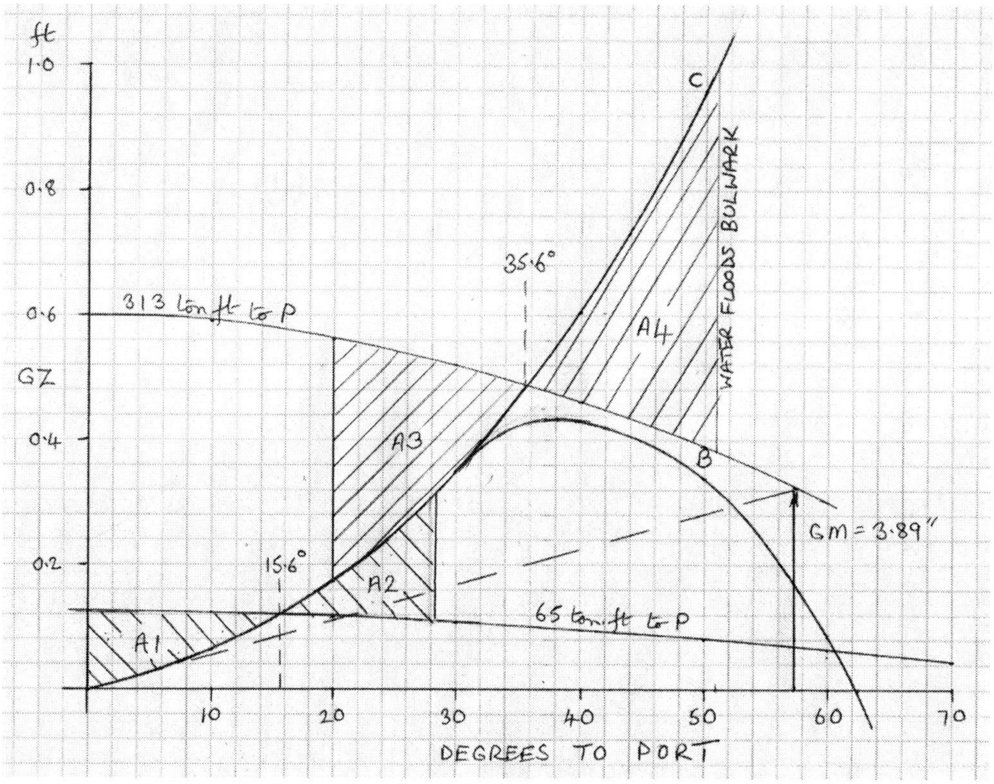

Figure 8: Reed's GZ Curve for *Daphne* redrawn to show effects of heeling moments

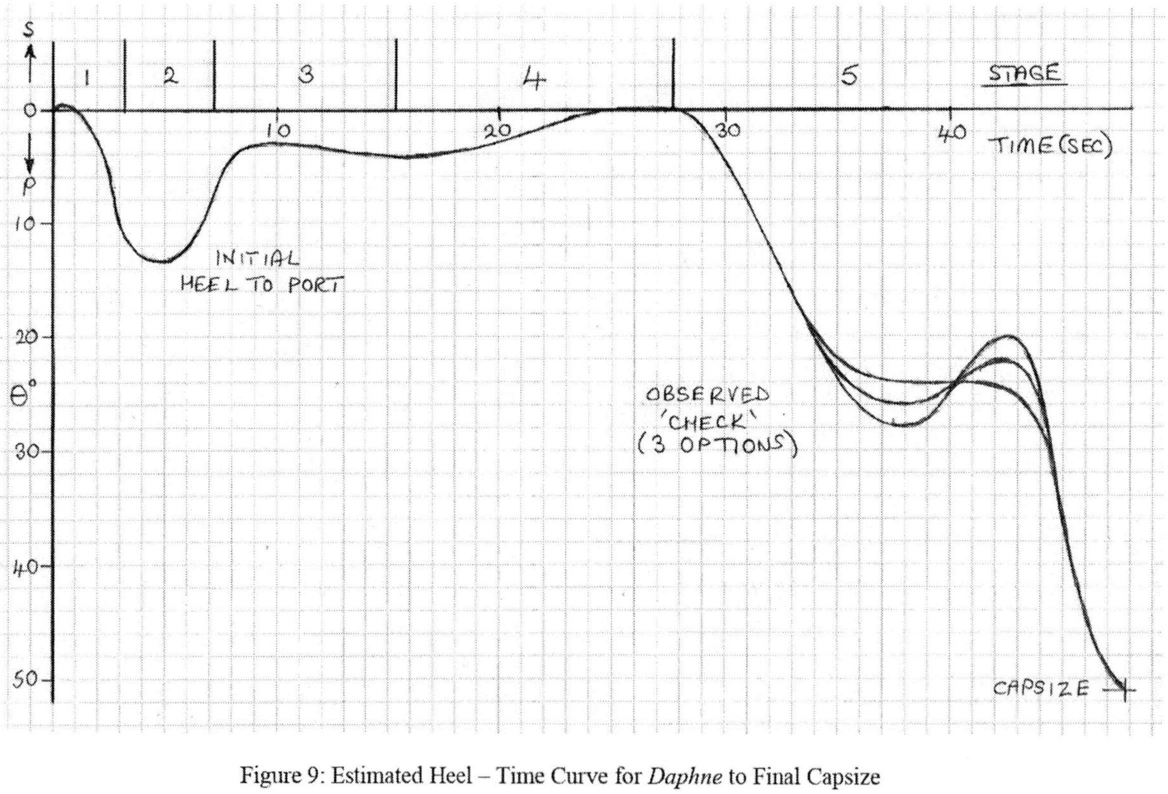

Figure 9: Estimated Heel – Time Curve for *Daphne* to Final Capsize

MEDIEVAL SHIPBUILDING IN THE ROMNEY MARSH

Fred M Walker, Fellow, UK

SUMMARY

During the fifteenth and sixteenth centuries, large sea going ships were built at Smallhythe – now a hamlet on the edge of the Romney Marsh. Probably 200 ships were constructed there in the same number of years including the great ship Jesus of 1,000 tons for Henry V in 1416 - then the second largest ship in England. The waterways around the Isle of Oxney have dried up but their positions are marked clearly by polder lands, and the marshes converted to sheep farming. Over a century ago Smallhythe lost access to the sea, and now the contact is the Cinque Port of Rye, eight miles south by road.

This paper will examine the shipbuilding industry of five hundred years ago, and comment on the particular problems of retaining skilled labour and in obtaining high quality raw materials. Sixteenth century shipbuilding administration and practice will be discussed and the contemporary ships (carracks, galleasses, balingers etc.) will be described briefly. Working in waters with a small range of tide and with little knowledge of ship theory was an added hindrance to efficient output.

For some years, Smallhythe was arguably the largest commercial shipbuilding complex in the south: It played a useful role when the Royal Dockyards were expanding and forced Henry Vlll to become awake to the risks of an "arms race" with Scotland and other European countries.

1. INTRODUCTION

Models of ships have hung in Churches since medieval times; often known as *Votive Models* they may express the gratitude of a family for the safe return of their dearest ones from a perilous voyage, or sometimes are memorials to a fine crew lost at sea. Denmark (where every Church is within 40 miles of the sea) has over 800 ship models, known there as *Kirkeskiber* similar to the Scottish name *Kirk Ships*. Since the Second World War, there has been renewed interest in Church Ships and many more models now are hanging throughout Britain – pride of place going to the lovely London Church of All Hallows-by-the-Tower, where about a dozen models are on display along with many other fascinating relics.

Visitors to the beautiful 12th century St Mildred's Church of Tenterden in Kent are attracted to an elegant and recently built model of the 450 ton galleasse *Grand Mistress*. This great ship – then one of the largest in the world - was built for King Henry VIII of England in 1545 and remarkably was constructed at Smallhythe, a tiny hamlet three miles south of Tenterden. Since the 16th century the sea has receded steadily and now five hundred years later the shipbuilding sites of Smallhythe, Reading Street and other places are flat polders supporting flocks of Romney sheep, and Tenterden is fifteen miles from the English Channel.

Some years ago, as interest in both Smallhythe shipbuilding and Church models was rising, a member of the congregation of St Mildred's Church offered to pay for a votive model of one of the "Smallhythe Fleet" to hang in the building as a reminder of our medieval shipbuilding tradition. The design was prepared by the author of this paper using "reverse engineering" techniques and then the model was constructed by a distinguished model maker on the staff of the National Maritime Museum, Greenwich. It now hangs at the west end of the nave of St Mildred's Church, Tenterden.

2. WHY WAS THE *GRAND MISTRESS* CHOSEN FOR THE VOTIVE MODEL?

This ship was chosen as she was one of the larger vessels built on the Romney Marshes and had been constructed at Smallhythe – which is within the Parish of St Mildred. Further considerable information was available in the Anthony Roll.

In 1546, a clerk in the Ordnance Office, rejoicing in the name of Anthony. Anthony produced a pictorial survey of 58 ships in the fleet of King Henry VIII. Each ship is shown in elevation with the starboard side in view, that is the ships are pointing to the right hand side of the page; almost all are drawn with flags flying, an anchor line extended and a pinnace or jolly boat attached at the stern. Ranking as one of the earliest public relations exercises for the Navy, the Roll endeared Anthony to the King Henry VIII who had by then realised the importance of naval power. One drawback of the Roll is that all the ships look alike, bringing into question the accuracy of the draughtsmanship. Any such problem is compensated for by the wealth of written detail regarding guns and equipment and by the listing of the complement carried broken into ***soldiours, marrynars and gonnars***. The drawings were on a single roll of vellum, which in the course of time was divided in three parts; the first and third now deposited in the Pepys Library of Magdalene College, Cambridge and the second or middle roll in the British Library, London.

3. BRITAIN'S ARMS RACE

It was fortunate that Smallhythe was well established when the King Henry VIII decided that a small fleet of ships should be assembled – all dedicated to his direct service, or in modern parlance - ships for the defence of the realm. Prior to that any invasion fleets or similar expeditions were dependent on suitable merchant ships being available for charter (being taken up from trade) or being requisitioned from dependable sources like the Cinque Ports as part of their contractual obligations to the Crown.

In the fifteenth century, most ships operating for the King were used as transports and the warship had not developed as a dedicated ship type. Very few warships had heavy guns mounted aboard, and their main offensive attributes were high towered castles both fore and aft which were manned by archers and by soldiers who were trained to leap from one ship to another during close encounters: Further most had platforms about half way up each mast, known as fighting tops where skilled marksmen and the best of the archers were positioned to pick off any enemy crew involved in hand to hand fighting on the open decks below.

On the death of King Henry VII in 1509, England had enjoyed peace for some years and the political landscape seemed reasonably settled with France. The thrones of England and Scotland were occupied by Henry VIII and James IV – brothers in law who were manipulating their relationships with great care as both royal families – the Tudors and the Stuarts had aspirations to inherit the British throne.

At that time, Scotland had a number of good ships and might be described as a small but efficient maritime nation. King James IV, a multilingual polymath, was not slow to appreciate how new methods of construction would enable larger and more efficient ships to be built – ships with wide decks, great structural strength and ballasted correctly to carry modern guns. A ship of 1,000 tons was built on the banks of the River Forth: Known as the *Michael*, she was the largest ship in the world and the most revolutionary, and all at a time when it was widely known that the gun foundry in Edinburgh Castle was carrying out experimental work on weapons for use at sea.

The new king of England – Henry VIII, until then little interested in political or scientific change, took notice that England was being left behind in an arms race dominated by other Europeans countries. He oversaw the build-up of his own fleet with ships ordered from the Thames, from the newly established Royal Dockyards and of course from Smallhythe. It is probable that the first deliveries of his reign from Smallhythe were the *Great Bark* and the *Lesser Bark*, both of 1515.

The two fledgling naval services of England and Scotland were the precursors of the later united British fleet that around the time of the Restoration (1660) was to be given a proud appellation: *The Royal Navy*.

4. WHY WAS THE ROMNEY MARSH USED AS A SHIPBUILDING SITE?

The four or five identified sites in the Romney Marsh now are all land-locked village, but despite this can claim to have carried out work for three English kings- Henry V, Henry VII and Henry VIII – indeed they can claim to have had a leading part in the founding of the Royal Navy. While readers may be aware that there are long lost shipyards in the south east corner of Kent, few realise that five hundred years ago shipbuilding – on a near industrial scale - was practised at Smallhythe, a parish of Tenterden. Indeed, if one excludes the Royal Dockyards, Smallhythe was the largest English shipbuilding site in the early 16th century.

Despite the fruits of increasing geographical and historical research, our knowledge of the Smallhythe shipyards remains limited, and any analysis can be little more than a tentative foray into their probable layout and administration. Information has been culled from a wide range of sources and we are fairly certain as to the position of at least two construction sites and also of the means of building, launching and ultimately moving of the "new-builds" from Smallhythe and nearby Reading Street down the River Rother to the sea. A conjectural portrait has been developed of the 15th and 16th century shipyards which will have to be tested in the light of ever improving historical and geomorphological information, and of our ever increasing understanding of medieval shipbuilding practice.

The evidence for shipbuilding at Smallhythe is conclusive: County records mention a wide range of industries which cluster round the ship construction sites including timber suppliers, animal haulage teams (the transport contractors of medieval times), charcoal burners, iron smelters, blacksmiths, rope and canvas makers, sail-makers and so on. Recent investigations have unearthed iron spikes and nails as used in shipbuilding in the Smallhythe area, and as an exercise, some of the roadways of 500 years ago have been walked over, checked and found suitable for the haulage of heavy lumber by horses and oxen. The most compelling evidence of all is from the records of the time of Kings Henry VII and Henry VIII which confirm that ships of the fledgling English Navy were constructed in the area. The Anthony List indicates several ships constructed at Smallhythe, Reading Street and Winchelsea: It is possible that some of the Winchelsea hulls were in fact built and launched at Smallhythe and then towed unfinished to Winchelsea for outfitting (that is the completion of the internal work) and for masting and rigging. Such a situation would arise when lack of water at Smallhythe forced the builders to keep the ships as light as possible enabling them to sail downstream with the least possible draft.

5. WHY WAS SMALLHYTHE SUITABLE FOR SHIPBUILDING IN MEDIEVAL TIMES?

In Medieval times, shipyards could be found in many parts of the coastline of Europe. Smallhythe was no different from the others and ship building there was subject to the following standard requirements:

Adequate Water – enabling ships to be launched safely, subsequently moored (preferably afloat) for the weeks or months while the finishing touches were bestowed on the new hull, and then with adequate water at high tide for the ship to leave for her appointed delivery voyage along the Rother Water and then southwards on the Appledore Water to Winchelsea and the sea.

Strong and Stable Land for the construction site - the ground which becomes compacted with regular use has to bear intense point loads from the ship under construction. It should have a gentle slope (known as the declivity) to the water's edge to ensure the completed hull can launch with the minimum of effort.

Storage Space - as shipbuilding is a complex assembly process, adequate space is vital, not just for the myriad of small items required in construction, but also for the vast quantities of lumber arriving on a daily basis from the forests. Once cut and prepared this material now known as dressed timber had to be stacked carefully to dry out partially before being used on the ship.

Access - was important for delivery of timber and other materials: In those days great reliance was placed on the Drove Roads – not only the "motorways" of the age, but also routes with the easiest gradients for the ox and horse drawn trailers in regular use.

Availability of Material – was vital: Delays in supply – always unacceptable – were overcome by holding adequate stock and by regular deliveries of items as diverse as rope and tar, iron pigs and lumber. Timber came in two forms, long tree trunks for planks and shaped pieces known as "grown timber" for the many small shaped items like knees, brackets, breasthooks and so on. The main timbers were:

Oak - a general purpose timber for keels, frames and shell planking and occasionally for decks.

Elm – a tough straight grained timber for keels and parts of ship subject to abrasion.

Fir – for masts and spars and sometimes decks and planking

Self-Sustaining – the community had to have inbuilt resources: A population large enough to maintain local services and supply labour and apprentices to the shipyards - housing to meet the needs of a thriving local community whilst accommodating influxes of itinerant labour in times of high production. In all respects Smallhythe came up to the mark!

6. THE WATERWAY THAT DISAPPEARED

Standing in the garden of Smallhythe Place and looking south over the low and fertile fields leading to the Isle of Oxney, one could be excused for doubting the existence of these medieval Kent shipyards, a doubt exacerbated by sight of the tiny stream at the bottom of the garden - the Reading Sewer, moving in a sluggish manner. The extreme flatness of the fields indicates they are polders, or land reclaimed from the sea, similar to the much larger areas found today in The Netherlands. It is not the remit of this paper to give scientific or detailed analyses of how the Smallhythe Channel to the sea changed to become a drainage sewer on the edge of reclaimed land. However a short simplified explanation follows:

The "shipbuilding era" of Smallhythe is assumed to have been from around the 1350s until just after the 1550s, and probably peaked in the 1400s. The numbers of ships constructed is unlikely to have been more than 250, many of which would have been local trading barges and small coastal fishing craft. Small vessels did not create much interest and few records exist.

For around two hundred years the valley between Smallhythe and the Isle of Oxney was navigable, and ships leaving Smallhythe (or Reading) would sail eastwards and on passing Appledore turn south in what may have been called Appledore Water. This channel seems to have been in fairly continuous use during the shipbuilding years, and at high tide the depth of water would have been adequate for ships bringing cargoes to Appledore, Reading Street and Smallhythe or other minor ports, or for vessels being delivered from the shipyards and on occasion returning for repairs. The twice daily ebb and flow of tides would have kept the channels clear of sand and silt, while any small shoals which might develop from time to time would be cleared by hand dredging.

Two explanations exist for the closure of the navigable channel, the first is a breach occurring in the dam at Maytham Wall sometime after 1600 which changed the shape of the channel, while the other was the deliberate breach of Knelle Dam in 1635 in a vain attempt to reinstate the navigable waterway. Around this time the shape of the navigable part of the Marshes changed dramatically with some parts becoming subject to flooding, whilst other parts drained. Smallhythe as a port diminished gradually and by the late 17th century, the ferry to the Isle of Oxney had been replaced by a bridge. The land profiles away from the channels will have altered little over the years, and therefore higher ground would look much the same as today. Similarly (but subject to further research) the sea levels are unlikely to have altered during the shipbuilding period by any more than a centimetre or two.

The actual depth of the waterways is unknown, but is unlikely to have been more than a few metres at high tide as shown in the diagram. Rather similar to Rye Harbour, at low water springs the Smallhythe Channel would have at most a couple of metres of water in the main channel, whilst at high water springs this will be several metres in depth. The difference is known as the Range of Tides and at Smallhythe the maximum range at Spring Tides is likely to have been no more than four and a half metres – and at Neap Tides a mere two metres. These estimates are based on our understanding that the maximum tidal range at the nearest part of the English Channel would have been nearer seven metres.

Subject to no barriers in the channel, ships with a draft of up to 3.5 metres could be moved from Smallhythe in the window of opportunity of about two hours on either side of high water springs. While tidal heights can be predicted with some certainty, from time to time (and usually owing to weather conditions) they fail to reach expectation and ships could be grounded on the passage and have to wait for as long as two weeks or one month until one of the following spring tides gave adequate water for open passage to the sea.

It is likely that shipbuilders would have kept the outfitting of their larger vessels to a minimum to ensure the displacement and also the draft was as small as possible. Some of the larger ships would not have armament placed aboard until delivered to Winchelsea and on occasions the masts and spars might not be shipped until the inland voyage was completed.

The ship movement would require oarsmen using long oars or sweeps working on the new ship as well as small attendant boats and they would travel as swiftly as possible, and with great luck make Winchelsea in one tide, otherwise the ship would be moored and might have to lie on the channel bed for some hours till the water returned.

There were no formal tidal predictions, and the shipwright in charge would have to work out suitable days for ship movements based on long experience of the river coupled with knowledge of the cycles of the moon.

7. ROBERT BRYGANDYNE – CLERK OF THE KING'S SHIPS

Formal records of the known two hundred years of ship building on the Rother are few and far between: We are struggling to identify less than twenty per cent of the ships built, and similarly are unable to name many of the residents of the community. Of the Master Shipwrights we know at least two of the family names - Hoggskynes and Brygandyne.

Hugh Roberts in his book *Tenterden: the first thousand years* relates that in June 1421, King Henry V on a visit to Smallhythe authorised a pension of 4d a day for John Hoggskynes, master carpenter "*because in labouring long about the ships he is much shaken and worsten of body*". Roberts suggests that this may have come through the monumental effort to complete Smallhythe's largest ship, the *Jesus* of 1,000 tons just five years before. King Henry VII is known to have visited Smallhythe in 1487, but according to Roberts, he made other visits to inspect the ships and is reputed to have had a friendly relationship with the shipbuilders.

The Brygandyne family was well known in Smallhythe with at least four related men who may all have served in the shipyards. The most prominent was Robert Brygandyne, a master shipwright in the 1480s and early 14890s, who in 1495 entered royal service as Henry VII's Clerk of the King's Ships. This senior appointment enabled him to oversee many important tasks including supervising the construction of England's first dry-dock at Portsmouth in 1496. At that time he was responsible for the building of the great ship *Henri Grace-a-Dieu* at Woolwich on the Thames: The *Great Harry* as she was nicknamed had her timbers cut in Smallhythe and transported by sea to Woolwich and further subcontracting included the refurbished of old anchors for the ship by the Blacksmiths in Smallhythe. In 1512 a large draft of shipwrights was assembled in the Tenterden area before they walked to Woolwich to work on the great ship; clearly Robert Brygandyne favoured the men of Smallhythe and used his considerable authority to ensure work came their way. Robert Brygandyne remained in his post until 1523 having served Henry VII and Henry VIII.

Royal patronage continued and when Henry VIII came to Smallhythe on 28 August 1537, he was received by John Brygandyne one of the following generation of Master Shipwrights: But by then the glory days were almost at an end.

8. THE MASTER SHIPWRIGHT

Always the *Key Man* in any shipyard, medieval or otherwise, the Master Shipwright had to ensure all work was carried out to the highest standard and that ships were delivered on time and at an agreed cost. Along with the privilege of absolute control of the shipyard, he had the ultimate responsibility of striving to make the business efficient and ensuring it played a worthwhile role in the relatively small community in which his family and friends made their livelihood.

The Master Shipwright was responsible for the selection of trees to be felled, a task that was critical to the success of the shipyard. Such work would take him on long and tough hikes through the forests of the South East to identify long trunks free of knots and in other cases timbers with unusual shapes (known as grown timber) that could be used for knees or brackets. His main task was the oversight of the shipyard, negotiations with contractors and ensuring there was suitable credit (hopefully in coin) to maintain steady employment. As one who had risen through the ranks, the Master Shipwright would know how to handle the normal problems of petty pilferage, insubordination, absenteeism, incompetence and drunkenness: However in the fifteenth and sixteenth centuries he was a powerful employer as there were few restraints regarding punishment and dismissal. However in a fragile local economy, unemployment was a serious matter and respect had to be shown by everyone.

It was usual for the Master Shipwright to lease the ground on which the ship was to be built, such as at Smallhythe around 1497 when the *Mary Fortune*'s keel was laid. The Master Shipwright was one of the Brygandyne families and the land rented from another member of the same family. Few shipyards had offices, and most used the local pub for meetings with customers and contractors – and on the very rare occasions when the workforce met as a group.

Apprentices were recruited from time to time, those with small ambition being little more than cheap labour, whilst others with drive and initiative were appreciative of this opportunity to learn some of the skills of the Master. Signed indentures bound them to stringent conditions for several years, and extracted a promise of non-disclosure to others regarding the "mysteries" of the Shipwrights' Trade. In a medieval industry where skill levels were not of the highest - mystique was a means of maintaining a good social and financial position.

9. THE SHIPBUILDERS

Up to the middle of the eighteenth century, the vast bulk of shipbuilders were itinerant workers who in those times of privation, moved along the coast between shipbuilding sites as work became available. Despite mostly being unlettered, they were proud of their skills as shipwrights and jealous of their rights to enjoy the relative freedom of working on ship hulls; they were "their own men". To this day groups of men continue to work in teams, often with long term links brought about by friendship and family connections, like the shipwrights from Australia, Denmark, Germany, New Zealand, North America and the U.K. who travel the world and can be found building replica sailing ships and other wooden vessels.

While it is impractical to build a sea going hull with a team of two men, a boy and a dog, equally it is remarkable just how much work can be completed by a couple of dozen motivated men in just a few months. The principal team members are **Shipwrights** (sometimes called **Carpenters**), men skilled in preparing, erecting and completing all the woodwork. They would work with their own small hand tools, kept sharp and in immaculate order and would be assisted by **Apprentices**, bound by indenture for something like five years, and working for derisory wages, and also by **Scavelmen** (semi-skilled waterfront workers) and **Labourers** (or helpers) for the heavy and less skilled parts of the work.

10. ANCILLARY WORKERS IN THE SHIPBUILDING ORGANISATION

The many and complex tasks in the shipbuilding process required many skilled ancillary workers. Some would be on the pay-roll of the Master Shipwright, whilst others would be working on a sub-contract basis. They included:

The **Sawyers**, whose task was to cut the raw lumber into planks or other required shapes for immediate use by the shipwrights. With the introduction in the twelfth and thirteenth centuries of efficient long two handed saws, the old trade of **Timber Cleaver** became redundant; the cleavers' task had been to split tree trunks lengthwise using axes, wedges and sledge hammers. This was a time consuming practice, but one that had been perfected by the Vikings a few hundred years before.

The **Spar Maker** whose task was to construct masts and spars for the new ship: Often not part of the shipyard organisation, but a person working closely with the Master Shipwright to ensure that the costly masts and spars were correct in size, shape and detail.

Another group allied to the shipwright's trade are the **Riggers**: Their task was to prepare and fit the standing rigging of the ship (the fixed stays and shrouds keeping the masts and bowsprit in place). Once the masts are positioned, the riggers arrange the running rigging – the halyards, sheets and lifts which enable the crew to work the ship. Closely associated with rigging is the work of the **Sailmaker**, usually an independent trader who purchased canvas in rolls for cutting and stitching to the required shape of sails.

Ropemakers and **Canvas Manufacturers** were essential trades, although owing to their high capital requirements, were found in urban areas and seldom in small shipbuilding towns.

The **Blacksmith** was an important person in shipbuilding communities like Smallhythe. At his forge, the fasteners for the hull planking – the nails, dumps and so on – were hand forged. The use of iron was developing in the 1400s and shipbuilding was enhanced with its use. Blacksmiths often worked on a sub-contract basis to the Master Shipwright.

Iron Founders in medieval time were dependent on supplies of iron ore and of charcoal. The iron used at Smallhythe came from smelters in various parts of the south of England, whilst the charcoal was manufactured by **Charcoal Burners** working in many parts of the Weald of Kent. Charcoal, rich in carbon, is the residue of timber heated within an enclosed earth dome and in near oxygen free conditions for about five days. Both the south of England and many parts of central Europe were de-afforested significantly by this process which continued well into the 20th century. Charcoal burners were never part of any shipbuilding organisation, but sold their product directly to smelters, blacksmiths and others.

The principal raw material for the shipyards was timber, and the suppliers were land owners and farmers in Kent and Sussex and possibly even further afield. Suitable shipbuilding timbers had to be chosen on site, felled, cleaned of branches before being transported to Smallhythe by ox or horse drawn waggons. The last group in the extended shipbuilding family are **Foresters** and **Drivers**.

The entire shipbuilding staff worked long hours, usually from dawn to sunset over a six day working week. Unless special arrangements had been agreed, most men worked for a daily sum of a few pennies. There were no first aid facilities and there was no compensation for accidents in the workplace which could be serious, if not fatal. Most injuries were caused by saws, axes, adzes and other hand tool, but also by falls or by objects falling from a height in the yard.

A health hazard singular to the area in earlier times was Marsh Fever, a form of Malaria carried by two types of mosquito found in Kent and East Sussex. After 1600 the disease became more prevalent when the amount of brackish or sea water declined, and an increased number of stagnant fresh water pools were able to host the mosquito larvae.

11. THE CUSTOMERS

The vast bulk of contracts at Smallhythe were for small vessels, and it is certain that orders for King's ships would not be forthcoming until the quality of work and standards of supervision had been established. Small fishing smacks, local ferries and small barges would have been part of their early work load and as time progressed more prestigious work came their way – including, in 1400 a barge for the town of New Romney, paid from civic funds. It is known that vessels came to Smallhythe for major rebuilds, while others on the river for trade used the facilities for voyage repairs. Payment systems must have been quite involved and would have included both straight financial transactions (possibly in different currencies) as well as part payment by goods and services. It is interesting to note that the East Sussex Record Office has some costings for small ship construction.

12. SHIP TYPES BUILT AT SMALLHYTHE

Ship types built in southern England five hundred years ago included:

Barges were the basic work-horses and seldom exceeded 50 tons. They had to be "worked" by sweeps, that is long oars, but many had one or two masts and could be sailed in suitable conditions. They were the precursor of the ubiquitous 19th century Thames Barge.

Balingers were vastly upgraded barges, with up to three masts and could carry cargo of up to 100 tonnes deadweight on continental voyages. The balinger was a near relative to the North European Cog, which was the main cargo ship used in the Hanseatic trade.

Pinnaces were small fast sailing craft, able to make short sea passages: Often they acted as tenders to larger ships such as galleasses. Most pinnaces had two masts and seldom were much longer than 14 metres on the waterline. One significant pinnace - the *Virginia* built in North America in 1608 to recognisable English design – is known to have crossed the Atlantic three times!

Great Ships also known as **Ships** and sometimes as **Carracks** were the largest vessels of up to 1,000 tons and with three or four masts. The two after masts, the Mizzen and the Bonaventure usually were rigged with lateen sails.

Galleasses were similar to Ships and in addition to their sails were equipped with long oars or sweeps: This enabled them to manoeuvre round pure sailing craft when in action. The working of the sweeps on the decks was found to be a hindrance and by 1550 all had been rebuilt as **Ships**.

13. SHIP DESIGN

Ship design has evolved over three thousand years - initially a gradual process relying on the empirical observations of master shipwrights and master mariners. Over the medieval period ships gained in size, their carrying capacity improved, they became just a little safer and shipwrights began to have some (very small) understanding of stability, or the means of keeping vessels upright. As ship plans did not exist before 1600, all the ships constructed at Smallhythe were "built by eye" under the close superintendence of the Master Shipwright, who would have his private guidance notes on the proportions of the hull cross sections and other matters – part of the mystery of the profession alluded to elsewhere.

From earliest times ship layout has been fairly standard irrespective of the country of construction. This was the result of the international dimensions of shipping, ship-handling and shipbuilding – businesses manned by personnel from every part of Europe. An example of this is the established order of securing rope and cordage on the deck enabling professional multi-national crews, despite their linguistic problems, to react promptly and in unison during an emergency.

14. CONSTRUCTION ON SITE

It is probable that two differing methods of construction have been used at Smallhythe. In the very earliest years a cradle forming the ship's shape was constructed into which the outer shell planks would have been laid (overlapping each other and in clinker format) and secured with nails or droves. Then the frames (the ribs of the ship) would be offered up and the hull formed. In early medieval times this system fell out of favour in Europe, despite remaining in use to this day in parts of South East Asia.

The majority (possibly all) Smallhythe ships were built using the frame first system, where a keel is laid down on building blocks. The keel is a key part of the structure and should be a single length of timber, but if short timbers only are available, the keel is fabricated and the parts bolted together at complex scarf joints. The ship frames are set vertically and at right angles to the keel line and then secured by the keelson, by the deck beams and by stringers. At this stage the ship is "in frame" and hull shape is now apparent. Incidentally frames are prefabricated using carefully sawn timber parts which are bolted together.

The hull or shell planking has to be prepared with great care as each plank is curved in two directions and has to be cut and bevelled before being subjected to heat in a steam box. Once reaching 100 Celsius the long planks are carried out on the shoulders of the building team, all protected by leathers and placed on the hull, bent to shape and secured. On cooling the timber sets permanently in its new shape. The hull integrity is completed with the laying of decks, formed of long planks pointing fore and aft and fitted such that the sheer (the fore and aft curvature) of the deck and the camber (the transverse round) of the deck is allowed for.

Being built of hundreds of component timber parts, the hull fasteners are items of great importance. Many are iron nails, droves and bolts, but the most intriguing are Treenails, often described as Trunnels which join large timbers together – see attached sketch.

Apart from hand tools the only major implements might be a saw-pit where two men worked double ended saws, the steam box of very simple design and a sheerlegs – a lifting device set up by the riggers using spars.

Ships were built bow to the water with the stern higher up. This allowed the strong stern post to be the point of attachment for the lines that would force the ship down greased wooden rails at the time of launch. Launching required considerable physical effort as it would be a couple of centuries before dynamic launching was evolved and ships slipped into the water under their own weight.

Once afloat, the ship would be ballasted with large smooth stones arranged to be supported on firm structure and they would be fastened to avoid movement in rough weather. The lower masts might be fitted and some of the outfitting which would include a stone galley hearth for cooking. Hopefully after a few short weeks the new ship would be part rowed and part towed to have heavy items like guns fitted, and there be prepared for the first sea voyage.

Building a ship is one of the world's most rewarding tasks, and an exciting part is the final handover of the New Rigged Ship, ready for sea!!

15. TONNAGE

Tonnage is a convenient method of describing the size - or the magnitude of a ship: It does **<u>not refer</u>** to the mass or weight of the ship, but rather the carrying capacity of the hull. In the 12th or 13th centuries, one of the major imports to our islands was wine from France and as this trade had to be regulated, it was necessary to estimate the capacity of ships. Around the 1420s, it was enacted that wine containers (known as ***tuns***) be of at least 252 gallons and from then on they were largely standardised and ships were known as having a capacity of so many tuns, a figure upon which port charges and taxes were paid. By sheer coincidence the mass or weight of a Tun of wine was close to the mass of one long (UK) ton of 2,240 lb.
As ship design was quite elementary, the authorities decided that for taxation and port due purposes ships would have to be measured for tonnage – confirming it as a capacity matter.

Until the 19th century ship tonnage was based on variations of the formula:

$$\text{Tonnage} = (L \times B \times D) / 94$$

Where L is the length of keel of the ship in feet
B is the breadth of the ship (inside planking) and again in feet
D is the depth of the ship from the top of the keel to the underside of deck at midships and once again in feet

While not perfect it gave a means for comparing the relative carrying capacities of different ships. For a quick estimation of tonnage, one may assume that the length of keel is between 5/8 and 11/16 of waterline length and that the L/B ratio is 3/1

16. SOME SHIPS KNOWN TO HAVE BEEN BUILT AT SMALLHYTHE AND READING STREET

1364	*La Gabriel*	barge?		
1378	*La Seint Marie*	barge?		
1394	*Marie*		100 tons	
1400	*St Eneswythe*	barge		for New Romney
1410	*Marie*	barge	100 tons	
1416	*Jesus*	ship	1,000 ton	
1416	*George*	ballinger	120 tons	
1486	*Regent*	ship	600 tons	
1497	*Mary Fortune*		80 tons	
1515	*Great Bark*			
1515	*Lesser Bark*			
1545	*The Grand Mistress*		500 tons	
1545	*Anne Gallant*	galleasse	450 tons	Built at Reading?
1546	*Great Gallyon*	ship	300 tons	
?	*Lesse Gallyon*	ship	200 ton	

17. AUTHOR'S BIOGRAPHY

Fred Walker served a Denny apprenticeship whilst studying engineering at Glasgow University. Following experience in Denmark, Ghana and the U.K., he was Shipyard Manager of Hall Russell for twelve years before becoming Consulting Naval Architect to the National Maritime Museum, Greenwich. Fred set up his own naval architecture business (in the ancient shipbuilding town of Tenterden) and has designed replica ships and overseen the restoration of many old vessels in many countries including Russia, the U.K. and the United States. As an established author, he lectures regularly at UCL (University College London).

Figure 1: A 500 year old terra-cotta coloured graffiti in Snargate Chuof of a "Great Ship" built a few miles away on Romney Marsh

Figure 2: The votive model of the *Grand Mistress* of 450 tons and built in 1545"

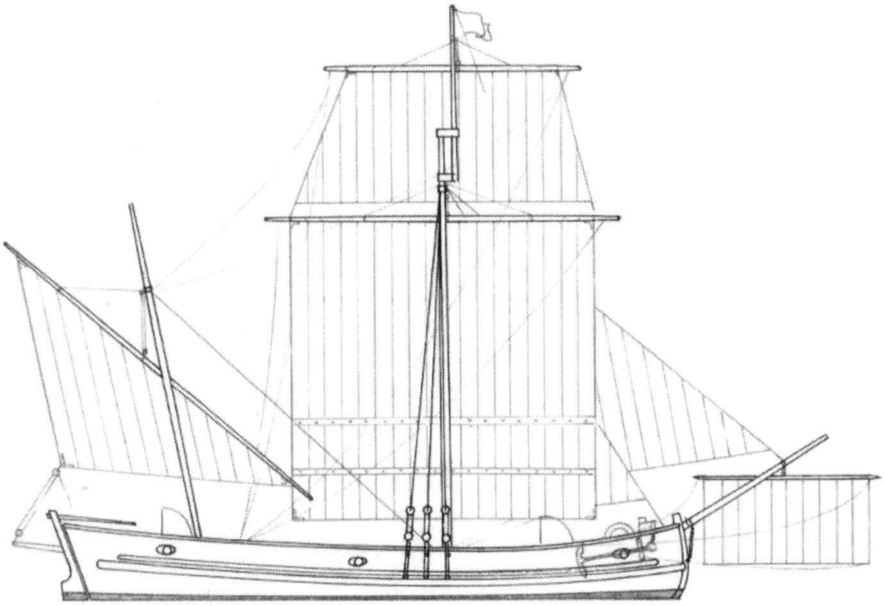

Figure 3: 16th century Pinnace

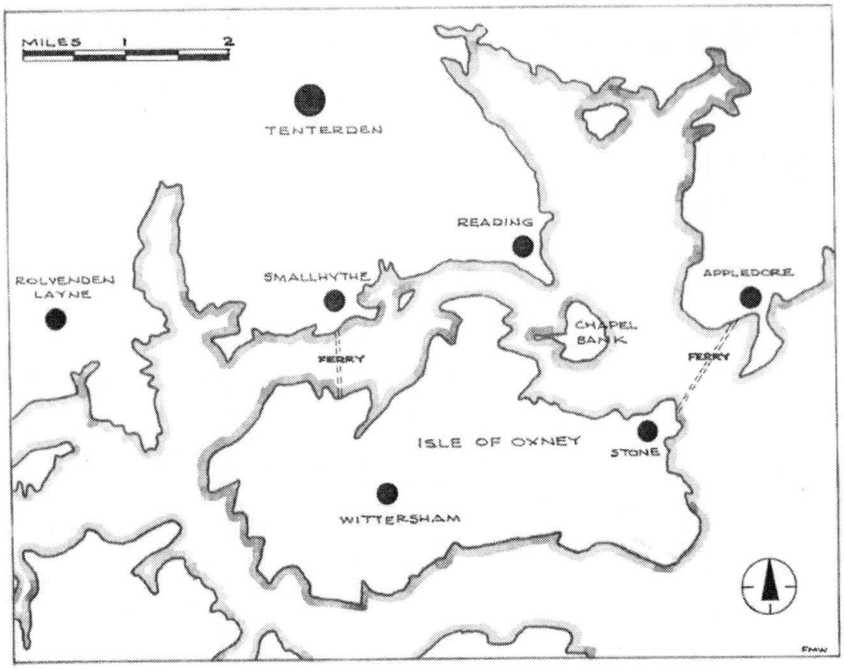

Figure 4: Map of the 40 square miles of Marsh land where ships were built

Figure 5: The Romney marsh at Fairfield

Figure 6: The Grand Mistress

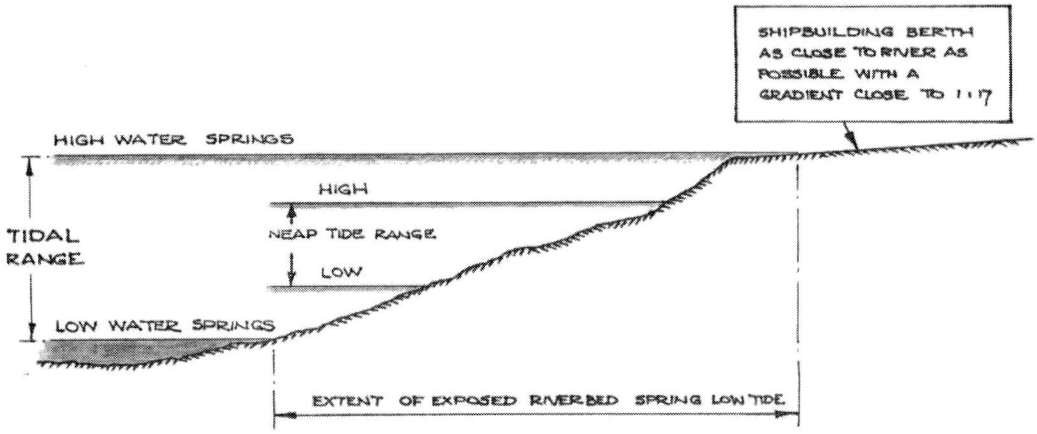

Figure 7: tidal range

TRANSITION

B.A.Mooiman, President Institution His Majesties "Bonaire", Netherlands.

SUMMARY

Like in other countries, Dutch 19th century shipbuilding underwent a fast development. Despite opposition from conservative naval officers Mr. Gerhard Moritz Roentgen, himself a naval officer is for a great deal responsible for the successful introduction of steam propulsion in the Royal Dutch Navy.

Half a century later Dr. B.J.Tideman became the most influential Dutch naval architect. Through the introduction of new production methods as a result of his own scientific research Tideman has restored the confidence in the Dutch shipbuilding.

He recorded his experiences in the "Memorial of the Navy", an internationally important book. His Majesties "Bonaire", built in 1877 was designed by Tideman as a product of the transition from sail to steam, from wood to iron and from muzzle loading guns to breach loading guns.

Of all Dutch warships built in the 19th century, His Majesties. "Bonaire" is the only preserved example.

1. INTRODUCTION

This is the short history of ship-building for the Royal Dutch Navy in the 19th century. In it the most important dates and names are given that marked a period of vast and fast technical revolution.

In the last quarter of the century a series of ships was designed and built for service in the Dutch colonies in the East and West Indies. One of this series is the ship, that is now under restoration at the old Royal Navy Dockyard Willemsoord in Den Helder, His Majesties "Bonaire". It is the only ship remaining from that series. For many reasons this ship is unique.

Figure 1: Zr.Ms."Bonaire"

In the future His Netherlands Majesties "Bonaire" will be visited by many interested persons who are willing to pay for their stay on board and in doing so contribute to the maintenance and therefore the preservation of this important piece of maritime cultural history.

2. SHIPBUILDING FOR THE ROYAL DUTCH NAVY IN THE 19TH CENTURY.

The 19th century was an age in which the progress of technology, also in shipbuilding and also in the Netherlands was enormous. In chronological order important dates in the development of Dutch shipbuilding in the 19th century are recorded in the following. Until around 1800 shipbuilding in the Netherlands hardly showed any innovative development. Ships were built like they always had been. Shipbuilders were not theoretically educated but relied mostly on experience. Besides, in Holland much use was made of the experience of English shipbuilders, and carpenters were often more influential than constructors.

In 1797, when Holland was a French republic the Dutch admiral in French service, Verhuel asked the French Minister of the Navy to place a small number of naval architect pupils at French dockyards to learn the trade. This was the first sign, that it was realised, that shipbuilding should take steps ahead instead of relying on the past only.

When Holland became a kingdom again after the fall of Napoleon, the Dutch navy possessed a number of ships of rather bad quality. The new king William 1 however was a great advocate of technical innovation. In 1815 he dismissed the Navy Constructor General, Mr. Glavimans because of the bad condition of the ships. Rumour had it, that in the French days, ships of bad quality were built on purpose. If these ships would be conquered by the English, at least they could not make use of them for a long time. Because of the French occupation from 1795 to 1814 contact with the rest of the world, but especially with England, was cut off.

In 1814 King William 1 started these contacts again and when in 1816 the English ships Defiance, the Orwell and the Caledonia visited the Netherlands the King showed great interest in these modern inventions and since then promoted the development of steam propulsion.

3. GERHARD MORITZ ROENTGEN

The name of Gerhard Moritz Roentgen became indissolubly attached to the introduction and development of steam propulsion in the Netherlands. Like in many navies, also in the Dutch Navy it took a long time before steam was introduced in warships.

Figure 2: Gerhard Moritz Roentgen.

Roentgen was born on May 7 in 1795. The Netherlands were so to speak a French colony named the Bataafse Republiek. Roentgen went to the Naval Institute in Medemblik to become a naval officer. In 1814 he served in the French navy in Toulon. With the fall of Napoleon he was imprisoned in the fort La Malgue in Toulon. But in the same year he succeeded in escaping to the Netherlands and here he was initially appointed sea-cadet but still in 1814 promoted to Lieutenant. In 1815, when he served on Zr.Ms. Brabant this ship was on her way to the East Indies but ran aground just outside the harbour of Den Helder. She continued her voyage but was leaking badly. The captain decided to enter Portsmouth and there Roentgen was introduced to English shipbuilding for the first time. Besides, some of the English warships he saw were propelled by steam engines. During his stay in Portsmouth Roentgen was tasked with the research in English ship constructions. Because of the many defects in Dutch ships, in 1818 Roentgen was again sent to England as a member of a committee to study steam propulsion the English way. On July 16 1821 he once more went to England to put out to contract and subsequently supervise the building of a steam vessel, later named "Wilhelmina".

In July 1822 Roentgen presented a memorial about the application of steam power on board warships. He is one of the founders of the Nederlandse Stoomboot Maatschappij (Dutch Steamboat Company) in Rotterdam of which he became a director in 1822.

Figure 3: Nederlandse Stoomboot Maatschappij in Rotterdam.

Because of the developments in foreign navies Roentgen was charged several times by the government to express his thoughts about steam propulsion for the Royal Dutch Navy. These thoughts were not always received favourably by the top of the navy. As a result of his frustrations Roentgen resigned in the beginning of 1824. Nevertheless he was installed as "Advisor in matters of Engineering".

On request of the King in 1824 Roentgen presented another memorial concerning steam power for the Navy. He advises to install steam engines in existing ships and to build new steam powered ships. These ships should be ram ships, should have less but heavier guns and should be protected by iron cladding. He even mentioned under water steam propulsion and guns. With this he was way ahead of his times and these recommendations practically had no follow up. In 1824 the King ordered an investigation into the feasibility and the advantages of a steamboat connection between Holland and Paramaribo, the capital of the then Dutch colony Suriname, a distance of 3900 sea miles. This investigation resulted in the purchase in 1826 of the Calpe, a steam paddler built in Dover. The ship was renamed Curacao. The ship weighed 436 tonnes and was bought for 11.500,00 pounds from the American and Colonial Steam Navigation Company. Even the engineer, Mr. Scrimshaw was taken over, who made himself famous the next year when the ship successfully crossed the Atlantic Ocean from Rotterdam to Surinam and back twice. Although more steam ships had crossed the ocean by then, the Curacao was the first ship to do so on steam power only.

Figure 4: Zr.Ms.Curacao.

In the following years many activities were taking place regarding steam shipbuilding. Not always with much success. Often Belgian machines, built by Cockerill were used but these showed many defects .The first steam ship, that had to be maintained by the Royal Navy was the Kings yacht, the iron paddle steamer "Leeuw". But the first Dutch warship was the already mentioned His Majesties Steam Packet Curacao.

The next steam ship for the Royal Dutch Navy, the steam packet Atlas was an outright failure. First it failed to leave the ways to be launched and when at last she floated her engines were breaking down repeatedly. The built of the second Dutch warship, the steam packet Suriname, was ordered in 1826 and completed in 1828. But also this ship suffered many defects in engines and boilers. After many repairs and a lot of money spent on her it was decided that the engines be put in another ship and the Suriname herself was demolished. The revolution in the Southern Netherlands, now Belgium and the consequent war between the Southern and Northern Netherlands resulted in the separation of the South from the North in 1830. This war exhausted the financial reserves of the Northern Netherlands. Therefore, unfortunately, the vast new-building program, started by the Dutch Government after 1815 had to be abandoned because of lack of money. Only some first class frigates, built according to French specifications were successfully built. The second class frigates, built in accordance with the principles developed by Sir William Symonds, surveyor of the Royal Navy were not successful and the building of more of them stopped.

Mr Roentgen's name is forever connected to the compound engine. Although Woolf's engine was already in use several years in this engine the pistons went up and down simultaneously. In Roentgen's engine however the pistons went in opposite directions, thus allowing the machine to be reversed any moment. Roentgens engine was first installed in the tug Hercules around 1829. Also in England confidence in steam propulsion was growing slowly but steadily, despite the lecture of dr. Lardner in 1835 in Liverpool in which he openly called the direct connection between Liverpool and New York by a steam boat: "perfectly chimerical". On the other hand around the same time Mr John Scott Russell wrote in: "On the nature, properties, and applications of steam and on steam navigation": "We are placed in that interesting epoch of time, when steam-navigation is an element in the destiny of the human race, more important than any art or invention of which we possess the records, if we except, it may be, the art of printing".

In the meantime the fame of Mr Roentgen's invention was spreading also abroad. In Rotterdam two complete machinery installations for French frigates of 540 nominal horsepower were delivered by the Netherlands Steamboat Company of Mr. Roentgen in 1840. This delivery was followed by that of two machinery installations of 300 nominal hp for two Russian frigates in 1842.

Despite the quick development of steam power in 1842 the Royal Dutch Navy possessed only 5 steamships; 3 in the Netherlands and 2 in East Indie. The minister for the Navy, Mr Rijk, wrote on July 18 1843 to king Wiliam 2 that the men, occupied with building ships were called chief carpenters and that that name did not give credit to these people who were often academically educated. He asked the king to establish the corps of the Naval Architects for the Royal Dutch Navy. Which happened.

1846 was an important year. A contract was arranged by the Navy with Mr. Roentgen's dockyard Nederlandse Stoomboot Maatschappij in Rotterdam for the construction of the "Semarang", equipped with steam engines of nominal 100 hp and for the first time in Dutch history fitted with an Archimedean propeller. The major part of the hull was made of iron, but the underwater part was composite with a double wooden hull.

The ship arrived in Batavia in 1848 and was a big success. Possibly the costs to build this ship were quit high. So high that many older naval officers were of the opinion, that steam warships were so expensive, that they should not be used in any war!

Proposals by the dockyard Nederlandse Stoomboot Maatschappij to install steam engines of nominal 300 hp and screws in the last two sailing frigates were not accepted. A decision which resulted in the Royal Netherlands Navy lagging behind in introducing screw propulsion in large vessels by approximately five years compared with other important navies.

Due to the breaking up of the Kingdom of the Netherlands in 1830 the country was nearly bankrupt as a result of which The Rotterdam Royal Navy Dockyard was closed in 1848 and the other Royal Dockyards were in poor condition.

Another result of the poor financial situation was the very slow increase in the number of steamships. In 1849 the Royal Dutch Navy counted only 12 steamships. But not only can the financial situation be blamed for this. Also the main sentiment in the Royal Dutch Navy was anti steam. Sailing ships were the real strength of the fleet, the admirals thought. Indeed there were some bad examples like the Atlas and the Suriname. And on top of that the exploitation of steam ships was very expensive.

Figure 5: Steam Packet "Atlas".

Still, in 1852 it was decided that all new warships were to be fitted with a steam plant albeit an auxiliary plant.
The engagement between Russian and Turkish warships in the Bay of Sinope in 1853 made it clear that new designs of warships were acutely necessary. Emperor Napoleon III initiated the design of a shallow-draught armoured vessel, suitable to deal with shore batteries and coastal fortresses. Because the French industry could not cope with the demand, the British Government agreed that 5 of these vessels would be built in the United Kingdom.

It became clear, that the tactics for naval wars were changing rapidly. The Dutch government realised, that the Dutch navy was in a difficult position because of a wide spread of tasks of her navy.

a. Warships as required by any navy
b. Vessels for colonial service, not suitable for other duties.
c. Vessels for inland defence in the Netherlands.

Because of this realisation Dutch naval constructors visited French naval dockyards in 1856 to study the floating battery "Devastation". The discussion started which type of ship could better defend the Netherlands and her colonies: the block ships or the floating batteries. Officers of the Dutch navy visited the Naval Review at Spithead on April 23 1856. Greatly impressed by the 4 floating batteries it was decided that floating batteries were preferable to block ships. Existing ships could be converted, using material which would come available when these vessels were razed and others scrapped. It was decided in 1857 that the "Koning der Nederlanden", originally the "Neptunis", the last three-decker of the Royal Dutch Navy, should be converted to a fully armoured floating battery called "Neptunus" again. The armour however was not included in the budget and would be provided later.

Figure 6: Zr.Ms."Neptunus"

In France, in 1858, the first iron cladded ship, "La Gloire" was constructed in Toulon to the design of the French constructor Dupuy de Lome. The English navy followed almost immediately with the "Warrior", launched in 1860.She was considered to be the world's first iron-hulled, sea going armoured ship and at her launch also the fastest and most powerful warship in the world.

Because of the rapidly deteriorating political situation in Europe a drastic change in design of the "Neptunus" was decided on in 1859. Her bow and stern were reshaped, 6 sixty-pounders fitted in each end of the ship, protected by 41/2 inch flat iron plates. A similar conversion of 6 additional vessels was executed.

But trial of "Neptunus" showed that:

- the unprotected sides were vulnerable,
- so was the position of the gunpowder rooms,
- the gun ports for the 60-pounders were too large,
- Smoothbore instead of rifled guns were fitted.

The many shortcomings of these ships were discussed in Parliament and as a kind of punishment the salaries of the Naval Constructors were frozen!!

On the whole, the transition of the former ships of the line into such floating batteries turned out to be a deplorable waste of money.

The introduction of steam in the Japanese Navy started in 1855 when the Dutch "Soembing" was presented to the Japanese emperor. As a result 3 more ships were ordered in the Netherlands by the Japanese. Also the education and maintenance of personnel and ships was completely provided by the Dutch Navy.

The battle between the Merrimac and the Monitor on March 9 1862 also had an effect on the Dutch ideas about the most effective warships.

Figure 7: Merrimac (left) and Monitor.

In the same year a committee advised the ministry of the navy, that 14 ships be converted into armoured steam batteries for the homeland and 2 screw corvettes into smaller batteries for the East Indies. The existing floating batteries were to be greatly improved. Some interesting designs were prepared by the Naval Constructor Bruno Joannes Tideman. Also in 1862 the Parliamentary Commission of Inquiry regarding the Navy suggested for the defence of the homeland: armoured gunboats of shallow draught, armoured floating batteries with minor steam propulsion and full steam batteries, suitable to operate in Dutch coastal waters.

A new committee, appointed to report on an improved Coast Defence System recommended in 1864:

For inshore defence:
- Twin screw armoured iron rams with a limited number of heavy guns.
- Twin screw, heavily armoured iron single turret monitors, armed with heavy guns.
- Floating batteries, using the existing vessels with minor improvements.
- Steam batteries of great speed and a draught allowing for leaving and entering the Marsdiep between Texel and Den Helder.
- Fast private steamers to be hired for reconnaissance.
- Existing gun vessels and gunboats.

After so many advises the minister for the Navy himself in 1865 ordered a twin screwed, twin turret, armoured vessel from Lairds at Birkenhead. The ship was state of the art and attracted much attention from the government, as the minister actually bypassed the chief naval constructor. The main reason for his action was the fact that Laird closely cooperated with captain Cowper Phipps Coles, the inventor of the revolving gun turrets. The turret ram was named: "Prins Hendrik der Nederlanden".

Figure 8: Zr.Ms."Prins Hendrik der Nederlanden".

Captain Coles was also the designer of the "Captain", the ship that in 1870 changed the way in which ships were designed. Froude and Rankine had already predicted that the stability of the Captain was reduced considerably above a certain heel. And indeed during a storm in the Bay of Biscay that was exactly what happened. As a result the ship capsized and 500 people were killed, the designer of the ship, Mr Coles, inclusive. After this disaster, the sailing rig on board steam ships was reduced to a mere cosmetic item.

Also the Dutch "Prins Hendrik der Nederlanden" showed the same dip in stability, and also on this ship the sailing rig was removed and the ship was banned to the East Indies.

4. DR. B.J.TIDEMAN.

From the 1870's the Dutch Navy made use of steam ships almost exclusively for the defence of the Netherlands and for the defence of the colonies sailing ships with auxiliary steam power were employed.

Also people started to realise, that the modern guns would always be able to penetrate the armour plating of the ironclads and that speed was a better defence than armour. From now on navies started to develop ships with high steam power.

But again, for the defence of the colonies still sailing ships, named cruisers, with auxiliary steam power were built. A very important reason for this was that for the long voyage the navy did not want to rely on unreliable bunkering stations.

In 1873 Tideman was appointed Chief engineer Adviser for Shipbuilding. In this function he became head of the Corps of Engineers of the Royal Dutch Navy, that was formed in 1843 and he was placed directly under the Minister for the Navy.

Figure 9: Bruno Johannes Tideman was born on August 7 1834 in Amsterdam and died there on February 11 1883.

In 1857 Tideman was posted at the Royal Dockyard in Flushing as an aspirant-engineer, after he had completed an education as shipbuilding engineer at the Military Academy in Breda and later, since 1863 at what became the Technical University of Delft. After paying some visits to England and France he was appointed to supervise the building of the first ironclad for the Royal Dutch Navy, HNMS "Prins Hendrik der Nederlanden" in Birkenhead. For this he stayed in England from 1865 till 1867 during which time he studied the technology and science in shipbuilding. In 1874 Tideman started a series of experiments at the Royal Dockyard in Amsterdam that brought him international fame. But many proposals he made for the Dutch navy were not executed because of lack of funds. It was Tideman's conviction that ship and propulsion unit should be designed as a whole and not separately. However he has not been able to convince the government of this. The ships were built at naval dockyards and the propulsion units at two private companies who shared the monopoly in this field.

Until 1874 the resistance of ships was calculated by simple and inaccurate methods. Rankine's formula was the most popular and it was derived from "perfectly shaped ships" so that the wave resistance could be neglected. This resulted in the evolution of fine ships with a large length to breadth ratio. This shape was undesirable for both merchant ships and warships leading to restriction in pay-load and weight for armament and armour plating. The formulation of laws to predict ships resistance by model tests was therefore an important step forward. This was possible by use of the law of similarity which had already been discovered in 1832 by the French naval constructor Reech and was rediscovered by the English civil engineer Wiliam Froude in 1869. It was now possible to determine the resistance of a whole variety of hull forms with the aid of models. William Froude showed by comparing model tests with the result of a full scale towing test of the destroyer Greyhound in 1874 that determining ship resistances by using models was reliable if one applied the law of similarity.

Figure 10: View of the trolley used to propel ship models across the first naval test tank constructed by William Froude.

Following this Tideman calculated the necessary power for the cruiser first class "Atjeh" by means of model tests according to the method of William Froude. The coefficients for calculation of the friction resistance Tideman derived from earlier publications by Froude on plank experiments. From this originated the well-known Tideman surface-friction constants. In 1876 Tideman established an experimental station in a sheltered part of the royal dockyard at Amsterdam, similar to Froude's tank in Torquay. However the method used for measuring the resistance of the models differed from Froude's method. A given resistance force was set up beforehand by using a balance and the speed of the towing truck driven by hand was adjusted to that of the model during the run. Thus only the speed of the towing truck had to be controlled.

Probably Tideman never had personal contact with William Froude although they had corresponded amongst other things about circular shaped armour plated ships. Tideman visited Torquay on 16 may 1879 where he was received by Robert Edmund, the son of William Froude who was expecting his father back from South Africa not knowing that Froude had died already on the 4th of May in Simonstown. Tideman also had great interest in steam engines and boilers. He did not agree that ships and engines should be treated separately by the Netherlands Navy. This is why he had no say in the monopoly position of the two large factories in the Netherlands , de Koninklijke Fabriek voor Stoom -en Andere Werktuigen in Amsterdam en de Nederlandse Stoomboot Maatschappij in Rotterdam which produced the steam boilers and engines for the Netherlands Navy. It followed from this that these factories were not in competition with British factories. Such competition was really necessary for after the Netherlands pioneer of steam navigation G.M.Roentgen had discovered the compound steam engine in about 1830 the developments were made possible by Elder in Glasgow in 1850 in obtaining higher steam pressures with an improvement in surface-condensation. The engines produced in the Netherlands were heavy and had high fuel consumption.

These circumstances, together with the Dutch government allowing the purchase by the subsidised steam navigation companies of packet ships from England due to the lack of confidence in their own shipbuilding industry caused Tideman to criticise sharply this behaviour of the government.

Tideman did much to show that the Netherlands were capable of building packet ships. He published an extensive book with designs called "Memoriaal van de Marine", which has been circulated widely abroad. He also made sure that the Netherlands were well represented at the Worshipful Company of Shipwrights Exhibition held at London in 1882.

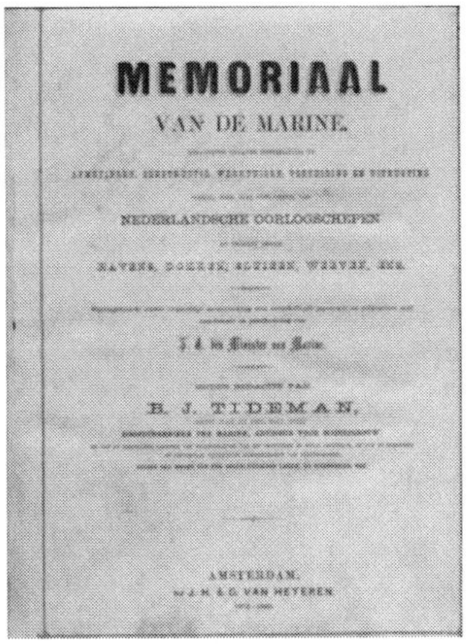
MEMORIAAL

VAN DE MARINE.

NEDERLANDSCHE OORLOGSCHEPEN

B. J. TIDEMAN,

AMSTERDAM,

Figure 11: Memoriaal van de Marine.

Already in 1872 cruisers 4th class were built. These ships were ordered by the Netherlands Navy primarily to serve in the West Indies and by the Department of Colonies to serve in the East Indies. The Netherlands Navy and the Colonial Navy were organised separately and each had its own fleet building programme. These cruisers were rather small ships of about 800 tonnes displacement, 54 meter long and 8, 4 meter broad. The ships were schooners with not so good sailing properties. They loofed badly. The Bonaire was one of this series and was built in 1876-1877. It is interesting, that although in his "Memoriaal voor de Marine" of 1876 he mentions that the guns would be by Armstrong, in the end Krupp guns were fitted. In these days the invention of breach-loading guns became well known and popular.

The story of the introduction of breach loading guns is also worth mentioning. In 1860 Alfred Krupp, the German owner of the enormous steel works, originally established in Essen, decided to make breach loading guns. But just like his earlier invention, the rifled barrel of his guns also the breach loader was considered by officers, in England and Germany to be a funny idea. (funny as a crutch). The German minister for defence, Mr Roon was not exactly a friend of Alfred Krupp. Only after England and France had ratified Krupps patent also Roon capitulated. Krupp delivered guns to Russia, Belgium, Holland, Spain, Switzerland, Austria and England. England was supposed to have tested Krupp's guns in 1862 at Woolwich and it had reported "extraordinary satisfaction owing to the complete tightness and safety of the breach mechanism". However, that could hardly be true, because in the mechanism a construction fault was incorporated that only showed up later in the war with Austria. But anyhow, England planned to acquire Krupp's guns. Alfred was well received in London, but the deal fell through because the emerging firm of Wiliam G. Armstrong turned the screws on Parliament. The Admiralty however secretly purchased Krupp barrels anyhow. Then, to Alfred's horror he discovered that Prussian admirals were eager to buy British guns. They wanted muzzle loading guns! Krupp complained to Bismark about this and Bismark agreed with him. Alfred's dominance at the 1862 exhibition was absolute. Alfred became known as the Cannon King. Then came the Bessemer process. Sir Henry Bessemer's convertor had been patented in England some years already. Krupp could have the Prussian license. But English ores, though not so pure as Sweden's, had relatively little phosphorus and were superior to the German ores. But in 1866 the war on Austria revealed the defects in the breach loading mechanism. Also in Russia a cannon had exploded during manoeuvres. Together with enormous financial problems Alfred broke down and disappeared for some time. Then Karl Wilhelm Siemens perfected the Bessemer steel making method achieving chemical changes in melted pig iron and steel scrap by burning gasses in the chamber. Although slower than the Bessemer process it produced far higher quality steel. Now, the German ores, so laced with phosphor, that they almost glowed in the dark, could also be used to produce high quality steel. Alfred Krupp took his chance and immediately came out of his voluntary exile.

In the war against Napoleon 3d in 1870 the Krupp guns literally smashed the French army and forced Napoleon to surrender. And as a result Krupp practically conquered the market worldwide, also the Dutch.

Since his appointment as Hoofdingenieur Adviseur voor Scheepsbouw Tideman was responsible for all designs, the Navy built. So also for these light cruisers.

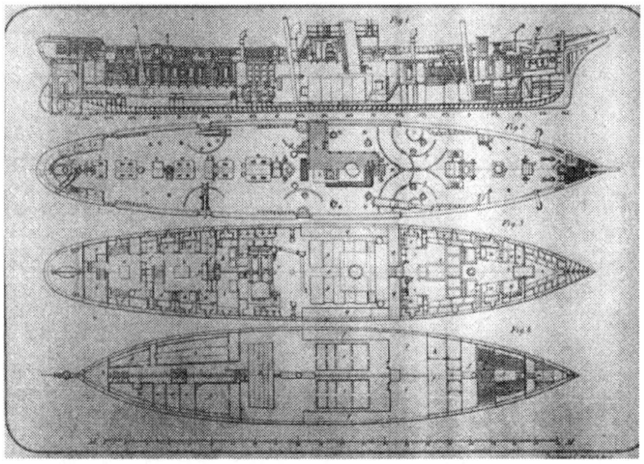

Figure 12: Zr.Ms."Bonaire"

During the design of steam ships Tideman still paid a lot of attention to the sailing-properties of the ships. He proved that by writing the book: "Verhandeling over de scheepsbouwkunde als wetenschap" in 1859. He emphasised the importance of rigging as broad as the length of the ship allowed. By doing so the sail-surface at a given heeling moment was as large as possible.

Another advantage was that a heeling ship with a broad and low rig had to give less rudder which gave less resistance and hence a higher speed. Because Tideman recognised the situation in England to be practically the same as in Holland he more or less translated the book, written by John Scott Russell in 1861: "The fleet of the future: iron or wood" into Dutch. In the book Scott Russell replies Sir Howard Douglass who speaks very negatively about iron ships and who unfortunately has great influence upon the policy of the Royal Navy. Scott Russell compares the properties of iron ships with those of wooden ships. In every respect the iron ship wins. Therefore he proposes to build a completely armoured iron fleet driven by steam only. He regularly mentions emperor Napoleon 3 as the great promotor of iron ships. He also mentions the Krim war in which ironclad ships have proven their big value.

But despite all promotion in the Netherlands for steamships still in 1882 the minister for the navy states that the best warship will be that which poses the best compromise between steam and sail while retaining its safety and seaworthiness.

When in 1888 Tidemans cruisers were compared with foreign warship during the fleet show in Barcelona, it was realised that these ships were practically too old-fashioned for warfare. The Royal Dutch Navy changed over to building modern steam ships that retained their masts but hardly had any sailing capability. Although still in 1892 it was written in the magazine "Onze Zeemacht" (Our Navy) that the rig was the only way for the Navy to be manned by strong, able men and not by children and old women. But in 1899 one of the members of parliament states: "when we compare the ships of our present war fleet with those of 12 years ago, we must, with regret of many of us, confess the total victory of steam over sail".

Perhaps as a result of the high number of difficult problems the politicians had to solve with regard to the Navy, the Dutch government counted 25 ministers for the navy from 1850 to 1900.

Figure 13: His Majesties Bonaire.

The keel of the Bonaire was laid September 21st 1876 in the dockyard of the Nederlandse Stoomboot Maatschappij, the Netherlands Steam Boat Company, started by Mr Roentgen in Schiedam. In his Memorial Mr Tideman mentions, that because the Ministry for the Navy had decided, that the draught should not be more than 3.5 meter he was forced to make the hull a bit fuller than he liked. For the same reason the ship has no keel but two floor head keels with a length of 20 meter. On May 12th 1877 the ship was launched and on September 25th it was delivered to the Dutch Navy and commissioned in the following November. At first this type of ship was called Schroefstoomschip der 4de klasse, Steam-propellorship of the 4th class. Later that name was changed to schooner. In fact the ship was a barkentine or schoonerbark with auxiliary steam propulsion. The screw was a Woodcroft-Mangin propeller that on earlier ships could be hoisted and stored in a recess in the rudderpost so that it did not drag in the water. On the Bonaire it could be hoisted over a smaller distance so that the ship did not always have to dock in case of repairing damage to the screw.

The most important sizes and technical particulars are shown in the appendix 1.

Although officially the crew was made up of 80 men, the still existing log registers a crew of 92 men. Obviously the number of crew members was dependent on the mission of the ship. The ship has been in active service from November 1877 until 1902. She had several tasks:

- She patrolled in the West Indies, primarily to protect the Dutch colonies against attacks by Venezuelan pirates.

- She protected Dutch nationals during wars or struggles in foreign countries.
- She made oceanographic expeditions and served in mapping unknown coasts.
- She supervised fishing in the North Sea and the Caribbean.
- Also representation was one of her tasks.

In 1898 a plan was made to change the ships engines to triple expansion engines to give the ship a higher speed. Also the boilers would be uprated to a higher capacity. That plan was not executed because of lack of money. Soon after the ship was decommissioned in 1902 and thoroughly rebuilt as a hotel ship for the School of Engineers of the Royal Netherlands Navy in Hellevoetsluis.

Figure 14: His Majesties "Bonaire" maimed as hotel ship in Hellevoetsluis.

In 1915 school and ship moved to Dordrecht from where in 1924 she was sold to the city of Delfzijl to serve as hotel ship for the School of Navigation Abel Tasman. The ship was also called Abel Tasman.

In 1988 the ship was not needed any more as such. People in Delfzijl have tried to use her as a museum but that effort failed. In 1997 she came to Den Helder to be inspected in the old dock, dating from 1822, of the Dockyard Willemsoord, that had just been vacated by the Royal Dutch Navy.

The condition of the ship was bad. For 100 years the ship had been used for housing the students of first the school of engineering of the Navy and later of the school of navigation in Delfzijl. Almost certainly the conditions of the iron hull and frames have hardly been inspected during that time. Condensation on one hand and the draining of washing water in the bilge on the other hand had caused quite a lot of corrosion. Also the 10 years of idleness in Delfzijl have not done the ship much good. The conclusion of the inspection was, that the ship could be restored, but at a price as low as possible retaining as much as possible of her originality.

Because of the very bad condition of the frames it was feared that, while docking the ship she would implode when the full weight would rest on the keel. Therefore a steel construction was attached against the hull so that 80% of the weight of the ship rested on this construction and 20% rested on the keel. In April 2004 the Bonaire had to leave the dock for some ceremony to take place on the dockyard and she was moored in the basin of the dockyard. Unfortunately the water level in the basin for some reason had to be lowered and the ship came to rest on the steel construction on portside only. The ship started to heel, all weight shifting to the portside steel construction that, as a result, was torn from the hull, leaving large tears in the plating. Within seconds the ship sank.

Figure 15: Bonaire submarine.

With enormous effort by divers and savers the holes in the hull were closed and a number of large pumps were able to suck the water from the ship. After lighting the ship, she was immediately towed into the dock again. If she would sink again, at least it would be a controlled plunge.

Figure 16: Repair of the hull. The poor state of the frames is clear.

Although the wreck looked terrible, the institution, meant to restore the ship was able to get a very generous subsidy from the City of Den Helder and started the restoration in May 2005. The speed of the restoration is completely dictated by the amount of money available. Some bursts occurred when this subsidy of the City of Den Helder was received. Although it was paid over a period of 10 years it showed the confidence and trust of the council that the restoration would be brought to a good end. Another enormous burst was caused by the subsidy of the Wadden Fonds. A fund helping touristic enterprises in the vicinity of the Wadden Zee. Also that money is now practically spent so it is time for another speedburst. But a fund, willing to provide a large subsidy has by now not been found. Then a subsidy was received from the Prins Bernhard Cultuur Fonds, a fund helping cultural projects. And although the amount of money was not as large as that from the other funds, the fact that this fund helped the restoration proves that the project is a recognised cultural project and this largely increases the status of the project. The result of all the investments is that the Bonaire gets back her original appearance and that the ship can be used by others for different purposes.

The ship will not sail any more. Also she will be dry-docked for the rest of her life. The reasons for that are many.

- A ship this size needs engines for leaving and entering the harbour. They will cost an enormous amount of money again for which almost certain no sponsors can be found.
- When under sail the ship needs a complement of at least some 50 to 60 men or women. A large organisation has to be set up to arrange all the necessary conditions.
- When in the water the beautiful lines of the ship's hull cannot be admired.
- Also the screw and the hoist to get the screw out of the water cannot be seen.
- But probably the most important reason is that the ship is docked in a graven dock dating from 1822 with a history, as interesting as that of the Bonaire herself. This combination is one to be celebrated.

Figure 17: Restoration underway!

5. THE CULTURAL HISTORIC VALUE OF THE BONAIRE

In 1876-1877 His Majesties screw-steam ship 4th class Bonaire was built at the Nederlandse Stoomboot Maatschappij te Fijenoord (Rotterdam) for the Department of the Navy. The ship was designed by Dr. B.J. Tideman, Hoofdingenieur Adviseur for Ship-building.

Mr. Tideman has in many ways put his mark on the Dutch, but certainly also on the ship-building in many other countries. Dutch dockyards had difficulties with the transition of wood to iron and from sail to steam. By introducing new production methods, developed from his own scientific research, Tideman restored the confidence in the county's own ship-building. He recorded his experiences in the "Memoriaal van de Marine", a book that was leading in the international ship-building.

This memorial has immense value for the restoration of the Bonaire. Many drawings of warships, among others also the Bonaire, and sections of the ships are clearly laid down and described. December 15 1977 was Memorial Day, held on the former Naval Dockyard in Amsterdam for the fact that the year 1877, also the year in which the Bonaire was completed, may be considered to be the year in which the activities in the field of ships model studies by Mr. Tideman reached their zenith. 5 years later a second "Tideman_day" was organised by the Maritime Research Institute of Wageningen and the Royal Dutch Navy. The screw steam ship Bonaire, designed by Mr. Tideman is a product of important, interesting and quick developments in the building of warships in the second half of the 19th century.

It was the era of transition from

- sail to steam
- wood to iron
- Muzzle loading guns to breach loading guns.

But also from dry compasses to liquid compasses,

Of all the Dutch ships, built in this technically revolutionary period only the Bonaire survived. Therefore the statement is true, that the Bonaire is an exponent of a very interesting period in time in naval ship-building and that she, where the construction of the hull in concerned is even unique in the world!

There are more reasons that the Bonaire should be preserved. The most prominent are:

- The Bonaire is the only Dutch built warship of the 19th century that is preserved.
- The Bonaire is the only remaining sailing warship of the Royal Dutch Navy.
- The Bonaire is the only remaining screw steam ship of the Royal Dutch Navy.
- The Bonaire is the only remaining ship in the world with an iron hull and a teak doubled hull.

When the Bonaire is restored she joins the 7 remaining monuments of the transition period from sail to steam and from wood to iron. The other ships are:

- HMS Warrior in Portsmouth,
- SS Great Britain in Bristol,
- SS Fram in Norway,
- The Jylland in Denmark
- The Uruguay in Buenas Aires in Argentine,
- HMS Gannet in Chatham,
- RRS Discovery in Dundee.

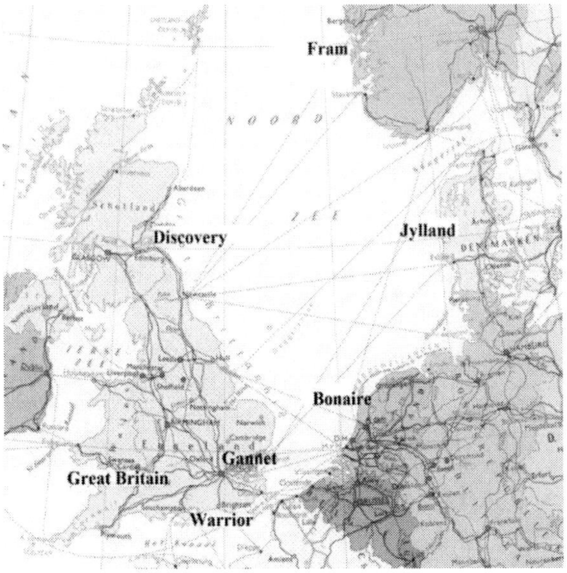

Figure 18: Places to go!

6. REFERENCES

1. M.G. de Boer, *Leven en bedrijf van Gerhard Moritz Roentgen*. 1923.

2. Bureau Maritieme Historie van de Marinestaf, *Aantekeningen betreffende de geschiedenis van het Korps Ingenieurs der Marine*. 1964.

3. E. Corlett, *The Iron Ship. The story of Brunel's ss Great Brit*ain. 1990.

4. J.M. Dirkzwager, *Dr.B.J.Tideman 1834-1883 Grondlegger van de mnoderne scheepsbouw in Nederland*. 1970.

5. W.F. Durand, *The resistance and propulsion of ships*. 1898.

6. A. van Dijk, *Voor Pampus. De ontwikkeling van de scheepsbouw bij de Koninklijke Marine omstreeks 1860*. 1987.

7. G.S. Emmersen, *S.S. Great Eastern, the greatest iron ship*.

8. G.S. Emmerson. *John Scott Russell. A great Victorian engineer and naval architect*. 1977.

9. J.F. Gugelot, *Scheepsbouw voor Stuurlieden en Machinisten*. 1952.

10. J.B.P. Haanappel, *De geschiedenis van de Marine-Stoomvaartdienst 1824-1949*. 1949.

11. F.W.K. de Klerk, *Klink los!*. Omstreeks 1950.

12. A. Lambert, *Warrior. Restoring the world's first ironclad*. 1987.

13. W.P.A. van Lammeren, *Het ontwerpen van scheepsvormen uit een oogpunt van weerstand en voortstuwing*. In Schip en Werf 2 en 3 februari 1940.

14. C. Maggs, *Isambard Kingdom Brunel. The life of an engineering genius*. 2016.

15. W. Manchester, *The arms of Krupp. The rise and fall of the industrial dynasty that armed Germany at war*. 1968.

16. B.A.Mooiman. *Het verhaal van de Bonaire*. 1999.

17. J.M.A. van Mulken, *Onze Gepantserde Schepen , beknopte beschrijving van het verband, de zamenstelling en de inrichting van het gepantserde materieel der Nederlandsche Marine*. 1874.

18. C.H.J. van Norden, *150 jaar Korps Officieren van de Technische Dienst*. 1995.

19. J.W. van Nouhuys, *De eerste Nederlandse transatlantische stoomvaart in 1827 van Zr.Ms. Stoompakket Curacao*. 1951.

20. F.C. van Oosten, *Schepen onder stoom* 1972

21. M.W.C. Oosterveld. *Tideman herdenking*. 1977.

22. J.S. Russell. *The fleet of the future: iron or wood*. 1861.

23. J.A.Schell, *Scheep-Stoomwerktuigkunde (naar A.E. Seaton)*. 1890.

24. A.E. Seaton, *The screw propeller*. 1909.

25. B.J. Tideman. *Memoriaal van de Marine*. 1880.

26. B.J. Tideman. *Het scheeps-stoomwerktuig van den tegenwoordigen tijd*. 1867.

27. Vereniging Oud-Oficieren van de Technische Dienst der Koninklijke Marine, *170 jaar onder stoom*. 1995.

28. A.J. Vermeulen, *De schepen van de Koninklijke Marine en die der Gouvernementsmarine 1814-1962*. 1962.

29. D. Vries, *Marine-ingenieur dr B.J. Tideman* . In Scheepshistorie 16 . 2013.

30. D. Vries, J. Klootwijk, J. Holtrop. *Schroefvoortstuwing, een 19de eeuwse technische uitdaging*.

31. In Scheepshistorie 17. 2014.

32. C. Young, *The incredible journey. The ss Great Britain Story 1970-2010*. 2010.

7. AUTHORS BIOGRAPHY

Ben Mooiman, commodore (E) rtd RNlNavy is president of the institution Stichting Zijner Majesteits "Bonaire" that restores the ship "Bonaire" at the old Royal Dockyard Willemsoord (1820) in Den Helder (Netherlands). Together with the only other board member he is responsible for all aspects concerning the restoration.

He has been the director of the dockyard from 1986 to 1993 before the yard was moved to the new Naval Base and was after that involved in the redevelopment of the abandoned site as a maritime cultural monument and tourist attraction.

APPENDIX.

Figure 19: Model of the Bonaire.

The most important characteristics of the Bonaire are:

Length over all	53,5 meter
Length between perpendiculars	45 meter
Greatest width	9 meter
Average draught	3,4 meter
Displacement	833 tonnes
Sail area	745,5 sq. Meter
Boilers	2 Scottish boilers of 65 lbs/sq.inch
Engines	2-cylinder horizontal direct acting steam engine
Capacity	412 indicator horse power
Propellor	4 blade Mangin propeller, hoistable
Ships speed	9,2 knots at 115 shaft revolutions per minute
Armament	1 breach loading gun 15 cm with rifled barrel
	3 breach loading guns 12 cm with rifled barrel
	2 mortars 12 cm
	2 revolving guns 3,7 cm

ENGINEERING ANALYSIS OF THE S.S. OHIO

C Bonnici, Department of Mechanical Engineering, Faculty of Engineering, University of Malta, Msida, Malta
C De Marco Muscat-Fenech, Department of Mechanical Engineering, Faculty of Engineering, University of Malta, Msida, Malta
R Ghirlando, Department of Mechanical Engineering, Faculty of Engineering, University of Malta, Msida, Malta

SUMMARY

The S.S. Ohio, that refused to sink under heavy bombardment, was towed to the Grand Harbour, Valletta, Malta on August 15th 1942 and saved the Maltese from capitulation during the Second World War. Historical literature, whilst focusing on her excessive bombardment, provides three main hypotheses as to why the T2-tanker did not sink under attack, namely as being: a strong fully welded hull, the density of cargo on-board and the use of water pumps to reduce the flooding rate. An analysis of the vessel's stability, floodable length and residual strength was conducted to prove or disprove the historical hypotheses. The results showed that the vessel was stable when damaged, was going to sink despite her welded construction and cargo onboard, but the water pumps did decrease the flooding rate. The research will be filling an existing vacuum in the area of knowledge, whilst providing concrete engineering evidence on how the S.S. Ohio survived.

NOMENCLATURE

AP	Aft Perpendicular
AVS	Angle of Vanishing Stability (°)
B	Moulded beam (m)
CoG	Centre of Gravity
D	Moulded depth (m)
GT	Gross Tonnage
GZ	Righting Lever (m)
L_{BP}	Length between perpendiculars (m)
LCB	Longitudinal Centre of Buoyancy (m)
LCG	Longitudinal Centre of Gravity (m)
LR	Lloyd's Register
TCG	Transverse Centre of Gravity (m)
VCG/KG	Vertical Centre of Gravity (m)
Δ_{LIGHT}	Lightship displacement (t)

1. INTRODUCTION

The 15th of August 1942 is the feast of St. Mary's Assumption (*Santa Marija* in Maltese), a feast which is fondly celebrated by the Maltese Catholic Church. However, on this very same day, 74 years ago, during the Second World War, the Maltese were also celebrating the arrival of the S.S. Ohio [1]. The T2-tanker was carrying vital oil replenishments, without which Malta would have capitulated to the Axis powers. En-route the tanker was heavily attacked, left dead in the water and had to be towed into Malta's Grand Harbour barely afloat but did not sink [1]. All those present were astonished on how she survived despite the inflicted damage. In view of this historic event, a research opportunity was identified. The unfortunate events surrounding the S.S. Ohio were studied and an engineering analysis was conducted to provide an engineering explanation of how the T2-tanker survived her ordeal.

2. HISTORICAL BACKGROUND AND HYPOTHESES

2.1 THE S.S. OHIO

The S.S. Ohio (Figure 1) was an American T2-tanker, owned by the Texas Company (later known as Chevron Corporation) and bareboat registered under the British flag during the Second World War [1], [2], [3], [4], [5], [6]. Her keel was laid at the *Sun Shipbuilding and Drydock* on September 7th 1939, four days after Britain and France declared war on Nazi Germany [1], [2].

Figure 1: S.S. Ohio [3]

Her length between perpendiculars (L_{BP}), moulded beam (B) and moulded depth (D) were 147.8m, 20.7m and 11.0m respectively [2], [7]. Her lightship displacement (Δ_{LIGHT}) was 5253t, with a 5438 GT and deadweight of 14377t [2]. Her ship side was riveted, whilst her nine centre tanks and twelve wing tanks were of welded construction, with a total cargo capacity of 1705m^3. The S.S. Ohio's main machinery consisted of two Westinghouse Steam Turbines, propelling the ship at a speed of 17 knots.

2.2 OPERATION PEDESTAL

The S.S. Ohio was part of the Second World War convoy, code named Operation Pedestal, later nicknamed by the Maltese as the Santa Marija Convoy. Operation

Pedestal was the last attempt to deliver desperately required supplies to the besieged islands, otherwise Malta would have surrendered within three weeks [1], [7], [8].

En-route to Malta the tanker was joined by its escort and 13 other merchant ships, carrying a wide variety of supplies [1], [7], [8]. However, as the sole oil tanker taking part in the convoy operation, only the S.S. Ohio was capable of delivering the large quantities of oil products Malta needed for power generation, food preparation and fuel to create an offensive against the enemy [1]. In fact, if the other 13 merchant ships of the convoy reached the islands whilst the S.S. Ohio was lost, the mission would have been a failure [1].

On August 12th 1942, the S.S. Ohio endured her first of many enemy attacks. Knowing how important the tanker was for both the Maltese and the British, the Axis Powers attacked her relentlessly throughout the convoy operation. So large was the inflicted damage that she was left dead in the water on the brink of sinking. She only reached Malta after being towed by two escorting destroyers from her port and starboard sides, into the Grand Harbour on the 15th of August 1942 (Figure 2). Her freeboard was at 0.76m (Figure 3) [1], [7], [8].

Figure 2: S.S. Ohio towed into harbour on the 15th of August 1942 [9]

Figure 3: S.S. Ohio with a freeboard of 0.76m, Grand Harbour, Valletta, Malta [10]

In the following years after the Santa Marija Convoy, the hypotheses on how the S.S. Ohio survived her ordeal were various, mainly being:

- auxiliary water pumps supplied by the escorting destroyers decreased the flooding rate [11], [12]
- the oil cargo being less dense than water prevented the ship from sinking [12]
- the strong fully welded hull was able to withstand such an ordeal [11].

Yet concrete engineering evidence, providing a clear explanation of how the S.S. Ohio survived her onslaught and disproving or proving the historical previous hypotheses does not exist. The main scope of the work was to establish the facts.

3. METHODOLOGY

A solution following the methodology listed below was developed for the research.

- First an information gathering exercise was undertaken to determine the main characteristics of the S.S. Ohio.
- A historical literature review followed to determine the chain of events surrounding the damage inflicted upon the S.S. Ohio.
- Next was the creation of a virtual 3D model of the S.S. Ohio using *Bentley Maxsurf 20.04 V8i Suite Software* which includes *Maxsurf Modeller Advanced* and *Maxsurf Stability Enterprise* [13].
- Afterwards the lightship centre of gravity was located.
- Finally stability, floodable length and residual strength analyses were carried out using manual calculations, results of which were confirmed using *Maxsurf Stability Enterprise*.

Through the adopted methodology and results obtained, valuable insight was shed on how the S.S. Ohio remained afloat.

4. CREATION OF THE VIRTUAL MODEL

A table of offsets is required to create a 3D virtual model of a ship. Unfortunately no such table could be found exists for the S.S. Ohio. Therefore using the method put forward by Bailey [14], the table of offsets was generated from a similar T2-tanker [15]. The table of offsets was inputted into *Maxsurf Modeller Advanced* to create the virtual model of the S.S. Ohio. Using a general arrangement plan provided by Lloyd's Register [16], the tanks and compartments were formed (Figure 4) and the margin line placed 76mm beneath the freeboard deck using *Maxsurf Stability Enterprise*.

Figure 4: S.S. Ohio 3D virtual model using

Next, the model was loaded with its crew mass, defensive armament to protect the ship against air attacks, required ammunition and full complement of cargo.

5. CENTRE OF GRAVITY LOCATION

A vessel's lightship Centre of Gravity (CoG) is of paramount importance especially when performing stability, floodable length and residual strength calculations. Unfortunately, the location of the lightship CoG of the S.S. Ohio was unknown and neither that of a sister vessel. Instead design methods and formulae implemented during the conception stages of ship design were utilised to reverse engineer the lightship CoG of the S.S. Ohio. The CoG is divided into three components:

- the Vertical Centre of Gravity (VCG)
- the Longitudinal Centre of Gravity (LCG)
- the Transverse Centre of Gravity (TCG)

5.1 THE VERTICAL CENTRE OF GRAVITY

The lightship VCG or KG was calculated using design formulae put forward by Schneekluth and Bertram [17], Ganesan [18] and Benford [19], who divided the lightship weight and VCG into three components:

- steel structure
- hull equipment and outfit
- machinery

The Principle of Moments was used to calculate the lightship VCG of the S.S. Ohio, which was located at 7.7m above the baseline after the weight and VCG of each of the above components were estimated using design formulae [17], [18], [19].

5.2 THE LONGITUDINAL CENTRE OF GRAVITY

The LCG of a ship lies above the Longitudinal Centre of Buoyancy (LCB). The latter corresponds to the centroid of the underwater area passing through the ship's centreline. However the LCB of the S.S. Ohio in her lightship condition was not available. Instead, knowing that when fully loaded the S.S. Ohio floated on an even keel [2], the LCB in the fully loaded condition was obtained using *Approximate Integration Methods*. Thus knowing the mass displacement of the S.S. Ohio in both the lightship and loaded condition, the LCB in the fully loaded condition (which coincides with LCG) and the mass and LCG of the cargo onboard, the lightship LCG of the S.S. Ohio was calculated using the Principle of Moments. The LCG was located at 70.8m from the AP.

5.3 THE TRANSVERSE CENTRE OF GRAVITY

The S.S. Ohio was symmetrical about the ship's centreline [3], [16], therefore the TCG lay on the ship's centreline.

6. DAMAGE SUSTAINED

En-route to Malta the S.S. Ohio was considerably damaged. However, for the scope of the analysis only the damage scenarios which caused flooding, as listed below, were considered. Any other damage which did not flood the tanker was excluded from the research.

6.1 DAMAGE SCENARIO 1 – 12TH OF AUGUST 1942, 19.56HRS

On the 12th of August 1942, at 19.56hrs, the tanker was torpedoed in her port side midship pump room, blowing a 24 foot (7.32m) wide by a 27 foot (8.23m) long hole (Figure 5) [1], [3], [12], [20], [21]. The torpedo explosion damaged four port wing tanks and two centre tanks [1], [3].

Figure 5: Torpedo damage [12]

6.2 DAMAGE SCENARIO 2 – 13TH OF AUGUST 1942, 08.10HRS

The following morning, August 13th 1942, at 08.10hrs, the S.S. Ohio was near missed by a 500lbs bomb. The force of the explosion punctured and flooded the ship's forepeak tank [1], [3], [12], [20], [23].

6.3 DAMAGE SCENARIO 3 – 13TH OF AUGUST 1942, 18.30HRS

Major damage was reported when on the same day at 18.30hrs, the S.S. Ohio received a direct hit in her engine room. A 500lbs bomb penetrated her boat deck, destroyed the engineer's accommodation on the upper deck and exploded in the boiler room, flooding the engine room [1], [3], [21], [22], [24], [25].

6.4 DAMAGE SCENARIO 4 – 14TH OF AUGUST 1942, 10.50HRS

Additional damage was inflicted on the 14th of August 1942. At 10.50hrs, an air attack developed over the S.S. Ohio [1]. A 1000lbs bomb was dropped in the immediate vicinity of the vessel's stem [1], [25]. The explosion flooded the aft peak tank and the steering gear room [1], [12], [20], [25].

7. STABILITY ASSESSMENT

A stability analysis was undertaken to assess the stability of the S.S. Ohio in both her intact and damaged conditions. First an equilibrium condition and a Righting Lever (GZ) curve (Graph 1) at the instant prior to the first damage inflicted on the S.S. Ohio were established using manual calculations. The results were verified using *Maxsurf Stability Enterprise*. The Lost Buoyancy Method was then utilised to establish the equilibrium condition of each damage scenario and using *Maxsurf Stability Enterprise*, the GZ curve of each damage scenario was plotted.

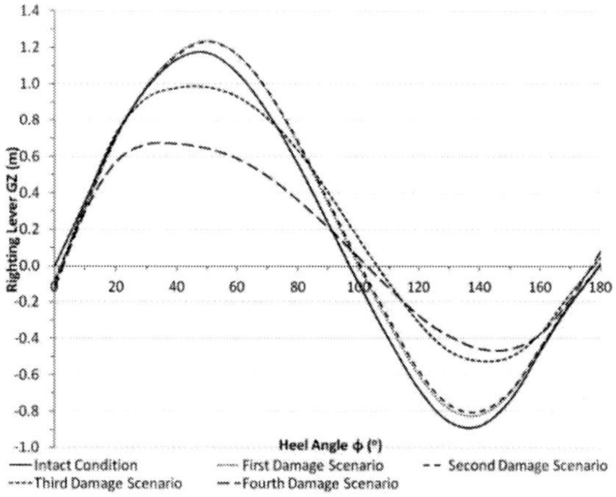

Graph 1: GZ curves for both intact and damage scenarios

The curves indicated that for the first two damage cases stability was enhanced, when compared to the intact condition. Reason being that a higher maximum GZ and a larger Angle of Vanishing Stability (AVS) were obtained for the first and second damage scenarios. Such a scenario occurs when there's either a reduction in displacement or an increase in freeboard. However due to cargo lost from damaged tanks and loss in buoyancy due to flooding, the S.S. Ohio experienced both a reduction in displacement and freeboard. Yet stability was improved because the reduction in displacement had a far more superior effect than the freeboard loss.

On the contrary, stability was decreased for the remaining two damage scenarios, even though there was a reduction in both freeboard and displacement. Such a scenario was attributed to the freeboard reduction being far more superior to the displacement loss. Otherwise the stability of the tanker should have improved rather than diminished with further damage. Furthermore the GZ curves showed that at neither instant of damage was the tanker in danger of capsizing, as the angle of final equilibrium of each damaged condition was less than 3°.

8. FLOODABLE LENGTH CALCULATIONS

Watertight subdivisions are of paramount importance to ensure that a ship is adequately protected against flooding. The floodable length calculations of the S.S. Ohio were based on the method put forward by Shirokauer in [26] as cited by Nickum in [27], results of which were verified using *Maxsurf Stability Enterprise*.

The floodable length calculations indicate the maximum length that could be flooded without immersing the margin line below the waterline and assuming the ship as lost [27]. The allowable flooded length is determined from the floodable length curve, which was plotted for the S.S. Ohio as shown in Graph 2. The curve indicates the maximum length, which could be flooded without immersing the margin line. A counteraction against flooding is to divide a ship into multiple watertight compartments. If a single compartment is flooded, the flooded length would be smaller than the allowable floodable length and the ship is not assumed lost [27]. In fact, the S.S. Ohio was divided into 18 transverse watertight compartments (indicated by 18 triangles on Graph 2), any of which could be flooded without immersing the margin line [3], [16]. As such the S.S. Ohio was a one compartment ship. However, if two adjacent compartments were flooded, then the flooded length would be greater than the allowable floodable length and the ship would be assumed lost. Yet despite being a one compartment ship, when the S.S. Ohio was torpedoed (Damage Scenario 1), three transverse compartments were flooded but she was not assumed lost. The floodable length of the breached compartments did not exceed the allowable floodable length and the margin line was not immersed beneath the waterline as shown in Graph 2. In fact, after the torpedo damage, she continued with her voyage under her own steam [1], [12], [20]. Similarly, in Damage Scenario 2, the margin line was not immersed beneath the waterline. The flooded length was still smaller than the allowable floodable length.

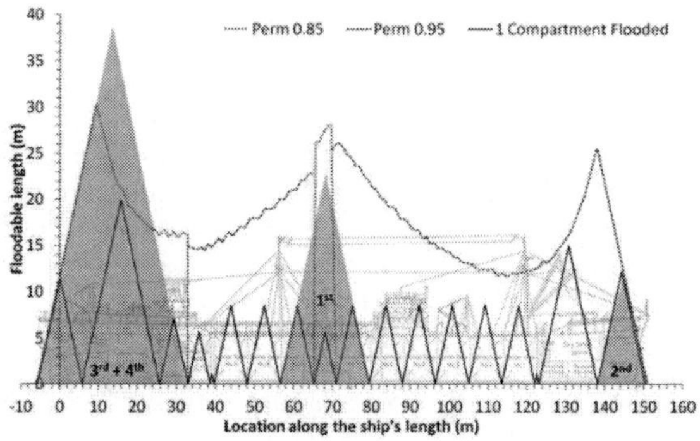

Graph 2: Floodable length curve of the S.S. Ohio with the flooded length of each damage scenario

It was not the case for Damage Scenarios 3 and 4. According to the floodable length curve, the flooded length of the last two damage scenarios was larger than the allowable floodable length with the margin line possibly immersed beneath the waterline. At this stage the S.S. Ohio was assumed lost and on the brink of sinking. The direct hit in the engine room and the near miss aft were the blows which could have sunk the tanker. History says that the S.S. Ohio was only saved due to the auxiliary water pumps brought onboard from the escorting destroyers to decrease the flooding rate [11], [12].

9. RESIDUAL STRENGTH ASSESSMENT

The residual strength assessment of the S.S. Ohio shed considerable light on the tanker's strength characteristics. Using Simple Beam Theory, the still and wave induced shear forces and bending moments acting on the tanker in both the intact and damaged conditions, were calculated. The wave heights required for the wave induced calculations were obtained from authentic synoptic charts (Figure 6) [28]. The results were then confirmed using *Maxsurf Stability Enterprise*.

The results showed that when fully loaded and in still water, the S.S. Ohio floated in a sagged condition (Graph 3). Thus the highest Bending Moments occurred when the vessel sailed through sagging waves.

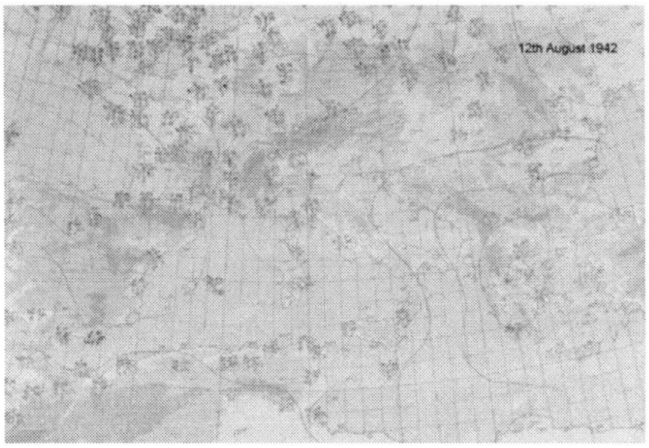

Figure 6: Synoptic charts of the Mediterranean as on the 12th of August 1942 from which the wave heights were obtained [28]

The method put forward by Paik, Thayamballi and Hong Yang, was also utilised to assess the vessel's strength when damaged [29]. Paik et al put forward mathematical relations to determine the Ultimate Bending Moment a ship could withstand when damaged before it collapses.

The residual strength results indicated that when the tanker was least damaged (Damage Scenario 1), she was experiencing the largest bending moment, which was greater than that obtained in the intact condition, as indicated in Graph 4. Against common perception the bending moment was decreasing with an increase in damage.

From the analysis, the maximum Bending Moment of the first damage scenario was equal to 588.4MNm, double the ultimate bending moment of 297.4MNm as predicted by Paik et al's method [29] for the first damage scenario.

As such, when the S.S. Ohio was torpedoed, she was at the greatest risk of hull collapse, despite being of a welded and riveted construction. As further damage was inflicted elsewhere along the ship's length, the maximum bending moment decreased. Reason behind the reduction in bending moment is due to the fact that the tanker's remaining buoyant sections were also flooded, decreasing her sagging condition, sinking the vessel on an even keel. Thus even though the S.S. Ohio was sinking further with each damage case, the damage itself prevented the ship from breaking in half.

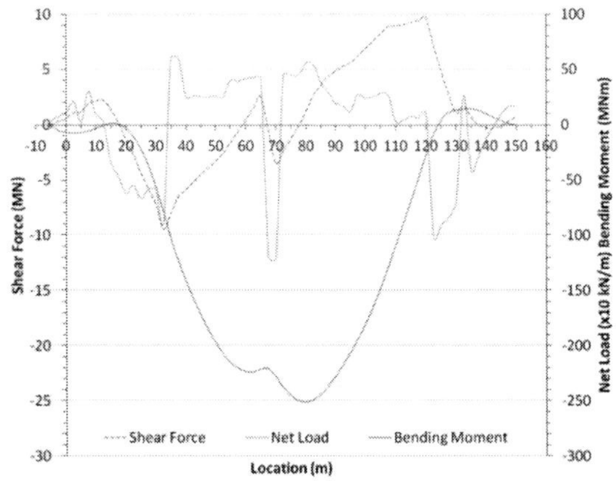

Graph 3: Still water net load, shear force and bending moment diagrams of the S.S. Ohio in the intact condition

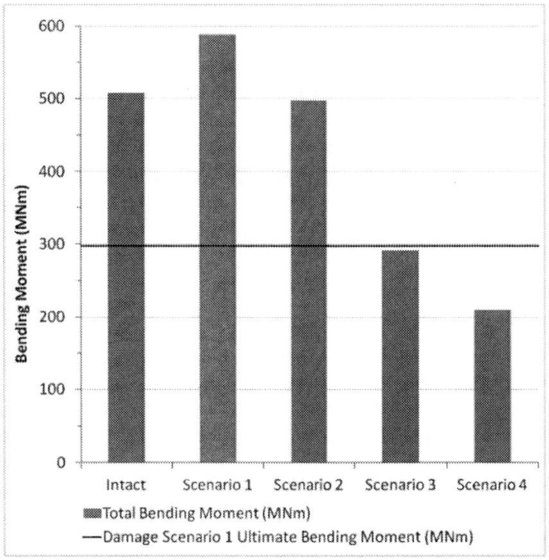

Graph 4: Bending moment for the intact and damage conditions

10. CONCLUSION

The stability, floodable length and residual strength analysis provided an engineering explanation of how the S.S. Ohio survived her ordeal.

The stability analysis indicated that despite the reduction in stability when severely damaged, the tanker was never exposed to a capsizing risk. The floodable length curve showed that when the engine room was flooded, the S.S. Ohio could be assumed lost but the water pumps decreased the flooding rate and provided ample time to reach harbour, proving the historical hypothesis. On the contrary, the floodable length calculations disproved the hypothesis that the tanker kept afloat because her oil cargo was less dense than water. The calculations showed that the tanker was assumed lost, even though the majority of the oil storage tanks remained intact. The residual strength assessment disproved the hypothesis that the strong fully welded hull was able to withstand her onslaught. When the S.S. Ohio was torpedoed, the maximum bending moment was greater than the Ultimate bending moment. At this stage the oil tanker was at risk of hull collapse despite being partially welded.

The research results presented here served as a knowledge gap closure within the historical field, since concrete engineering evidence was provided on how the tanker withstood her ordeal. The analysis will join up some pieces of the long outstanding puzzle of how the S.S. Ohio survived during the Second World War.

11. ACKNOWLEDGEMENTS

Acknowledgements go to:

- Allan Shaw - S.S. Ohio veteran
- Anne Cowne - LR
- Antonio Prestigiacomo - LR
- Bertin Rodolphe - STX Europe
- Emmanuel Magro Conti - Malta Maritime Museum Curator
- Gian Luca Mantegazza – RINA
- Howard Clayton - ABS
- Joep Bollerman - LR
- Joep Broekhuijsen - Damen Schelde Naval Shipbuilding
- Joerg Poettgen - Meyer Werft
- John Harper - Chevron Corporation
- Jörg Peschmann - DNVGL
- Joseph Stephen Bonnano - U-Boat Navigator
- Kim E. Demory - USCG
- Kurt Guttridge - Naval Architecture Services
- Leslie Schembri - ABS
- Marcus J. Akins - USCG
- Mark Beswick - UK Met Office National Meteorological Archive
- Michael Pearson - Author of *The Ohio and Malta – The Legendary Tanker that Refused to Die*
- Neville Galea Pirotta - Geography & Creative Art Teacher, St. Michael's School
- Patrick Camilleri - S.S. Ohio aesthetic model builder
- Patrick Chaurand - STX Europe
- Paul Caruana - Historian
- Rosalie van Egmond - Shell International B.V.
- Simon Cusens – Historian

12. REFERENCES

1. M. PEARSON, *The Ohio and Malta: The Legendary Tanker that Refused to Die*. England: Pen and Sword Books, 2004.

2. THE TEXAS COMPANY, "1939-1940 S.S. Ohio Sun Shipbuilding Construction Reports," Chevron Corporation, California, USA, 1940.

3. C. H. BOGART, "Operation Pedestal," *Steamboat Bill,* vol. LI, No.3, pp. 173-185, 1994.

4. THE TEXAS COMPANY, "Trial Trip of a Texaco Tanker," *The Texaco Star,* vol. 27, pp. 14-17, 1940.

5. UNITED STATES COAST GUARD, "Ohio Registration Card - ON: 239732," 1942, June 2.

6. UK MINISTRY OF WAR TRANSPORT, "Transcript of Register for Transmission to Registrar General of Shipping and Seamen," 1942, June 29.

7. L. LUCAS, *Malta - The Thorn in Rommel's Side.* UK: Penguin Group, 1993.

8. J. HOLLAND, *Fortress Malta.* UK: Orion Books Ltd., 2003.

9. SGT. DEAKIN, "Operation Pedestal and the Siege of Malta, August 1942," *Imperial War Museum,* vol. IWM GM 1505.

10. H. E. COOK, "Operation Pedestal and the Siege of Malta, August 1942," *Imperial War Museum,* vol. IWM GM 1480.

11. P. SHANKLAND AND A. HUNTER, *Malta Convoy.* UK: Fontana Press, 1983.

12. J. CARUANA, "Ohio Ordeal," *The Sunday Times,* 1992.

13. BENTLEY SYSTEMS MAXSURF 20.04 V8I SUITE SOFTWARE, (2014, January 6) "Maxsurf - Naval Architecture Software for all Types of Vessels", https://www.bentley.com/~/asset/14/2708.ashx [Accessed 2014, February 10]

14. D. BAILEY, *The NPL High Speed Round Bilge Displacement Hull Series: Resistance, Propulsion, Manoeuvring and Seakeeping*

Data. Royal Institution of Naval Architects, 1976.

15. UNITED STATES MARITIME ADMINISTRATION, "Historic American Engineering Record – USS Saugatuck T2-SE-A1," *HAER,* vol. HAER No. VA-128, 2006.

16. LLOYD'S REGISTER, *S.S. Ohio Scantlings and General Arrangement Plan.* Royal Museums Greenwich, London, UK.

17. H. SCHNEEKLUTH AND V. BERTRAM, "Chapter 5: Computation of Weights and Centres of Mass," in *Ship Design for Efficiency and Economy*, 2nd ed. Oxford, UK: Butterworth-Heinemann, 1998, pp. 149.

18. V. GANESAN, "A Model for Multidisciplinary Optimization of Containerships," Virginia Polytechnic Institute and State University, July, 1999.

19. H. BENFORD, "The Practical Application of Economics to Merchant Ship Design," *Journal of Marine Technology,* 1967, January.

20. P. SMITH, *Pedestal - the Convoy that Saved Malta.* England: Crecy Publishing Limited, 1970

21. ADMIRALTY, "Ohio War Loss Cards," Guildhall Library, London, UK, 1942-1946.

22. J. WYLD, "Report by Chief Engineer Mr. J. Wyld," Shell Historical Archives, The Hague (SLA), SC 88/5 file 1.

23. D. E. BARTON, "Operation Pedestal - M/011281/42," London National Archives - ADM1/15526, United Kingdom, 1942, August 20.

24. D. W. MASON, "Report by Captain Dudley. W. Mason," Shell Historical Archives, The Hague (SLA), SC 88/5 file 1.

25. H. J. A. S. JEROME, "Report on Operation "Statue" from the evening of Wednesday, 12th to 0800 of Saturday, 15th of August," London National Archives - ADM1/15526, United Kingdom, 1942, August 26.

26. F. SHIROKAUER, *A Simplified Method for Exact Arithmetical Determination of Bulkhead Curves.* 1928.

27. G. C. NICKUM, "Subdivision and Damage Stability," in *Stability and Strength*, 1st ed., E. V. Lewis, Ed. USA: The Society of Naval Architects and Marine Engineers, 1988, pp. 111-194.

28. MET OFFICE NATIONAL METEOROLOGICAL ARCHIVE, *Climatological Return for Valletta for August 1942 and Northern Atlantic Charts which Cover the Mediterranean for 1st to 15th of August 1942.* Exeter, United Kingdom.

29. J. K. PAIK, A. K. Thayamballi and Soo Hong Yang, "Residual Strength Assessment of Ships After Collision and Grounding," *Mar. Technol. SNAME News,* vol. 35, pp. 38, Jan 1998.

13. AUTHORS BIOGRAPHY

Colin Bonnici holds the current position of Flag State Inspector within the Technical Department at Transport Malta, Merchant Shipping Directorate which is the Maltese Flag Administration. He graduated as a mechanical engineer from the University of Malta in 2012 and graduated Master of Science in Engineering in 2016.

Claire De Marco Muscat-Fenech holds the current position of Associate Professor with the Mechanical Engineering Department at the Faculty of Engineering University of Malta. She graduated as a mechanical engineer from the University of Malta in 1989 and graduated Doctor of Philosophy in Engineering in 1992 from the University of Reading.

Robert Ghirlando holds the current position of Professor with the Mechanical Engineering Department at the Faculty of Engineering University of Malta. He graduated BSc(Eng), University of Malta in 1968, MEng, University of Liverpool, 1969 and PhD, University of Liverpool, 1974.

EARLY CROSS RIVER VEHICLE FERRIES

A D Moyse BSc, CEng. MRINA, MIOSH Retd., UK

SUMMARY

As Europe developed in the late middle ages it became necessary for wheeled vehicles to cross rivers in increasing numbers. Sailing ships had obvious limitations as had paddle and pole propelled vessels.

Horse Propulsion was not very successful in Europe but in USA these vessels became an important part of the transport system.

Europeans were successful in harnessing the fast flowing rivers by anchoring midstream and tacking from bank to bank. In this paper these vessels are referred to as Pendulum Ferries.

The invention of steam propulsion and reliable chain in the early 18th Century caused some Pendulum Ferries to use the High Wire system (to facilitate Chain Boats) and also Floating Bridges.

While horse propulsion has completely disappeared, ground chain and wire guided ferries have a useful role in todays transport system. Bridges or tunnels may ultimately replace most of these ferries if local cross river traffic volumes increase substantially.

1. INTRODUCTION

When only a farmers cart or the Lord of the Manor coach and horses needed to cross rivers a set of oars man on a fine day was sufficient. When roads developed and several coaches with two or three sets of horses arrived at crossings, ferries developed.

Four developments or inventions occurred: in the late middle ages:
- paddle wheels powered by horses
- vessels harnessing river flow
- steel wire
- chains hauled by steam engines.

I started my working life at the Vosper Ltd, Camber shipyard next to the then Portsmouth Harbour chain ferry. From The Point one could see the MOD Director Class paddle tugs and the Isle of Wight paddle steamers. It was the end of era.

Chapter 4 resulted from finding a wreck report [9]. In the transition from Sail to Steam there was an interesting development of Horse propelled ferries that is unknown about in Europe.

My interest in European inland waterway began with a study of 'The Operation of Coasters into the European Inland Waterway' [1]. It was while working in the Netherlands I came across a Gierpont and in German Gier Faehre. These river propelled ferries are described as Pendulum Ferries in Chapter 5 and their subsequent development as High Wire Ferries in Chapter 6.

Research into these and wire guided ferries naturally led me back to the UK chains ferries which as a boy I had seen at Portsmouth and Southampton which are discussed on Chapter 7.

Ferries in this paper are considered to be Vehicle vessels unless prefixed by 'pass' for passenger/cargo.

Chapter 2 considers the sites and environmental factors that these vessels have to work within.

2 SITE INFLUENCES

2.1 THE RIVER

Rivers have a consistent direction of flow and seasonal fluctuations in height. Too high could cause flooding on approach roads and loading difficulties due to adverse inclination of vessel ramps, too low could cause similar loading difficulties, negligible under keel clearance with possible damage to rudders etc.

Those rivers sited nearer the sea will have tidal influence with reversal of direction as well. A maximum significant wave height of 1.2m would be expected under modern European Union regulations.

Only short distances between landings could be envisaged with oars; ropes, wire or chain guided vessels would increase this, and self-propelled vessels would increase this even more as long as there was enough power to battle against wind and current

The latter permitted landings to be positioned in the most advantages positions rather than immediately opposite each other.

The ferry is a continuation of the road each side of the river so gentle gradients are required. Any excessive gradient shore side or on the vessel itself is undesirable especially in wet and icy condition.

Slipways have the advantage of coping with variation of river height best but require end loading vessels

2.2 SLIPWAYS

An early river vessel propelled by Oars is shown below.

Figure 1: Ferry at Worms about 1645

It can be used for this chapter as an example of an early end loading vehicle ferry which has just arrived from rowing directly across a river [2] [3].

As the river height changes so does the bow of the vessel move up or down the slipway. Stability of the vessel is not a problem as long as the carriage and horses remain still. There are no mooring lines shown so the river is presumed slow running. It is unclear from the picture if the vessel ends had ramps but the carriages had large diameter wheels which are an advantage for uneven surfaces.
Oars and man power will be discussed in chapter 3. For just one four wheeled coach and two horses there are five men, four for the side oars and one at the stern for steering. Larger vessel would have required even more man power.

This basic arrangement has become the norm for most ferries with mooring arrangements being located on the sides leaving clear deck space for tandem parking.

2.3 QUAYS

Vessels loading at quays ideally require a floating platform level with the vessel deck edge. Excessive rise of fall of the river can produce steep side ramps which is undesirable. Single hulls loading over the side would need to be wide enough to ensure heeling was acceptable.

Twin hulls vessels are inherently more stable than single hulls. The space required for horses and carriages is limited by the side bulwark entrances which are located near amidships and the vehicle turning arrangements once on board anchoring and mooring arrangements also affect the useful length of deck as does the tiller.

2.4 NAVIGATION ASPECTS

Besides manoeuvring around other shipping, sand banks also have to be avoided when crossing rivers. Some vessels may need to turn and berth or unload onto a slipway which is parallel to the river. Prevailing winds may affect this.

The weather and landings on opposite sides of any river may be different because of fields next to the river bank or vertical rocks.

Any cross river wire connections need to be sited above mast height and either be easily disconnected or heavy enough to sink to the channel bottom.

3. ENERGY SOURCES

Prior to the industrial revolution to cross a river required the assistance of one of the following sources

- Wind
- River flow
- Men
- Animals

The energy required would be for a short distance, that between arrival and departure points. We will look at these in general terms leaving the application to vehicle ferries to subsequent chapters.

3.1 WIND

Mariners have used sails from early times .The Wind is unpredictable: it does not blow at the right times or in right direction, it gusts and sometimes does not blow at all.

Sails on small ships are normally rigged fore and aft leaving little space for side access between masts for deck cargoes. Rigging also restricts on deck space. Such design constraints limited their usefulness for ferries.

3.2 RIVER FLOW

Just as wind produces a propulsion force other than dead ahead, so moving water with the aid of a rudder or an unsymmetrical orientated hull gives a transverse force.

For slow flowing rivers and lakes Man or Animals were adequate except in windy conditions. Vessels using fast flowing river water appear to have been used mainly in Germany and the Netherlands as we shall see in Chapter 5.

High water and low water can cause restrictions not least because of current speed but complete cessation is rare except in winter when rivers can and do freeze.

3.3 MAN

Men and horses have been used to tow vessels *along* rivers and canal banks for centuries. A history of Dutch experience can be found in [3] It is interesting to note that that the towage rope here and *along* the Rhine [4] was often attached to the stern via a high pennant rope controlled from a forward mast, presumable to clear bank obstructions and limit deterioration by rotting.

The article [3] also shows a picture of an Amsterdam horse propelled ferry of 1827 which proved to be an American design and is discussed in Chapter 4.
Pole propulsion requires the pole to touch bottom so is limited to shallow rivers and lakes. Both wind and current speed can restrict usage to very local application. There is a famous picture in the UK showing the Horse Ferry which operated between Victoria and Lambeth. It shows twin horses and a carriage on deck and a Pole man!

We saw in chapter 2 a vessel propelled by four oarsmen and a steersman. Clearly a larger vessel would require larger numbers of oars men or some devices that could coordinate and enhance their working together.

Capstans were used on sailing ships to weigh anchors and assist with rigging from early times so their usage together with water and wind mill technology was inevitable.

The problem of how the energy of mens arm and leg muscles could best be used for propulsion began in the 17th century.

3.3(a) Side devices

Leonardo da Vinci is credited with a drawing showing a small boat with side paddles joined together by a cranked shaft operated by a man. [5]

Reference [6] dated 1698 suggests that if eight oars could be mounted on a drum head [P&S] on the end of a transvers shaft then a capstan with crew could be used to drive them. See Figure 2.

One of the interesting things about this diagram is the capstan idea. Clearly 4 or more men could work together. This may be a key stage in the development of external paddles but as we shall see in Chapter 4 these ideas had actually come together in 1680.

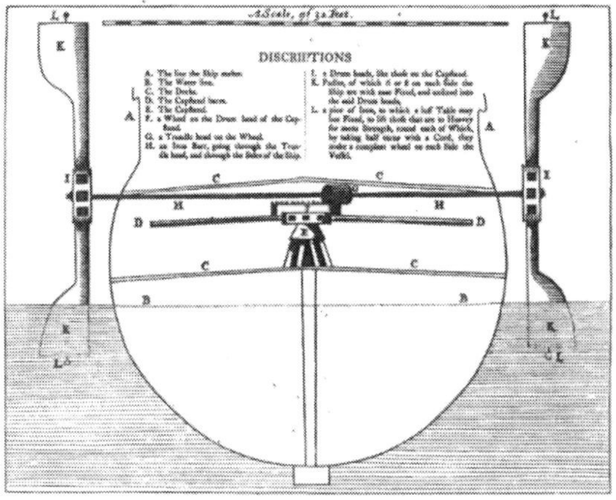

Figure 2 Saverys Oars Paddle

3.3(b) Bottom Devices

According to [7] Nicholas Witsen (1641 - 1717) built the first ship in the Netherland which was partially assisted by human muscle. The article also mentions a Patent granted to Lucas van Baten and sons Lucas and John in January 1663 for propulsion wheel with hollow spoons

In the Amsterdam Ship Museum there is a model of such a ship - Figure 3

Figure 3 Model MC 1013

This figure shows a horizontal crank in a compartment next to the shell which drives a wheel inboard which is in a longitudinal cofferdam. This wheel drives two other wheels either side fore and aft which have duck feet around the perimeter. These two duck wheels can be raised or lowered as necessary. When lowered the duck feet protrude below the hull.

This is an *internal* paddle, presumably with one on each side. It is not clear whether it is Foot or Hand driven but the space appears wide enough for one person.

3.3(c) Stern Devices

Paddles at the stern seem to been considered only from the beginning of steam propulsion.

The early days of submarine design shows man power had a limited application there though with screw propellers.
- the Confederate submarine CSS Hunley of 1863 had an in line hand crank system for 8 men to the
- the first German submarine which was designed in 1850 was driven by two separate tread wheels

There was only one way of increasing power and that was to turn to the farm animals, the horse being the power source.

3.4 ANIMALS

The larger farm Animals have clearly more energy than a man. In Romans times oxen and mules had been considered for the propulsion of vessels but it is not known if any were built.

Horses being more plentiful in Europe were therefore the obvious animal to use being capable of working all day and all week.

The transmission system was the problem. A whim required a diameter of perhaps 3 times the length of the animal so any vessel deck had to be at least10m wide. When this was tried the horse suffered from giddiness. A device enabling the animal to be stationery but use its legs to create motion was required.

This development occurred in the USA as will be seen in chapter 4.

4. HORSE PROPELLED FERRIES.

4.1 CONCEPT

I stumbled across a report on the Burlington Bay Horse Ferry Wreck [9] which operated on Lake Champaign, Vermont, USA in the 18th century and found horse-propelled vehicle ferries had existed there just before steam propulsion started. This development in the USA is virtually unknown about in Europe.

Paddles developed with horse ferries initially but the maximum power available was only 8HP which was apparently adequate for American Rivers. A greater power source was soon to be developed.

4.2 PADDLES

We saw in Chapter 3 that man power was limited but farmers had used their animals for other uses. Horse towing *across* rivers, being pulled across the river from both sides alternately, is referred to some documents in Europe.

The concept of paddle wheels driven by animals certainly goes back to Roman times. But it is not known if they were actually built. [10] shows a triple whim/capstan driven by Oxen with six side paddles.

Europe

In the late Middle Ages horses were used for sawing wood and dredging. Water and wind mill technology would have been relevant to the paddle revolution.

The earliest known successful horse propelled vessel in northern Europe is believed to have been designed by a cousin of King Charles 11, Rupert, prince palatine of the Rhine in 1680. [11]. It is thought to be a horse whim design for 8 horses and could out pace a royal barge with sixteen oarsmen. It or something similar was used in Chatham Dockyard in 1682 and was said to be capable of towing the largest ships of the Royal Navy. Little is known about these vessels.

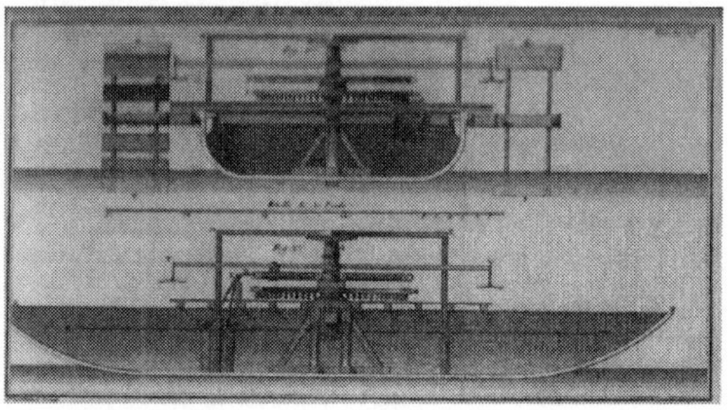

Figure 4 Compte de Saxe

The diagram in Figure 4 was published in 1732 [15] and is attributed to Maurice, Compte de Saxe. It was approved by the French Academy of Science. It has similarities to Saverys proposal but it was never built.

Reference [7] says that in 1743 a ship named Volg My [called Malle Schip, meaning Sillyship by the locals] was built for the Zaandam to Alkmaar service in the Netherlands. .During its first and only voyage it began to sink and five horses appeared and swam away but another one drowned. One drawing show a completely enclosed deck and another just a deck house. Neither show side paddles so bottom scoops as in Figure 3 might be assumed.

Reference [7] also refer to two vessels built in the UK
- 'The Safety' built on the Clyde in 1812 for service between Liverpool and Runcorn was converted to a horse paddle boat in 1818
- 'The Courier' built in 1814 as a steam paddle boat for service between Norwich and Great Yarmouth. A sister vessel exploded so The Courier' was converted to horse powered paddle wheel for passenger confidence reasons .These vessels were for passenger and cargo as far as is known.

However Rendel's first UK floating bridge [Chapter 7] was converted back from steam to horse whim in 1836 which indicates horse machinery was being developed.

Reference [3] part 2 shows a horse propelled paddle ferry with single horse and carriage on the open deck. The deck ends are rounded and no ramps are indicated suggesting side loading. The depth of the hull above water is greater than the height of a man.

It was ordered in 1827 by the Amsterdam City authorities and was one of two built to an American design! The vessels were built by Van Vlissingen & Dudok van Heel. The first went on trial 25 June 1829 and showed it was under powered for the conditions. The second was launched in 1832 and was taken out of service 1858.

Further information is given in [12] and [13]. The horses for this design are situated below deck on a tread mill which may account for its large hull depth, there is no reference to underdeck arrangements. The drawings do not indicate how the horses entered the engine room from the deck.

It is said the design was under powered and this might be confirmed by an incident in 1833 when strong winds and currents caused drifting off course. It is also claimed that a Jetty needed adjustment for its use.

America

While developments in Europe were slow they took off on the East Coast of N. America. The first successful horse propelled vessel is believed to be in 1791 on the Delaware River and was based on a Horse Whim or Capstan. It had twin hulls with a paddle and whim on the centre line amidships. Such a passenger vessel is shown in Figure 5.

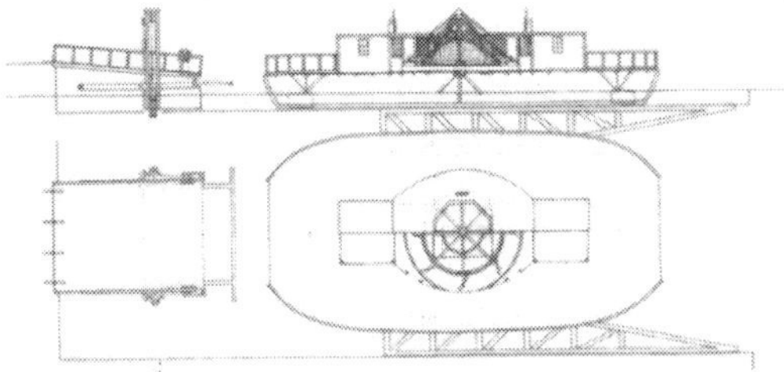

Figure 5 Horse whim ferry

From 1807-10 there was a horse propelled passenger vessel called 'The Experiment'. It was propelled by a 'goose foot paddle' which was kind of propeller [14] It was driven by eight horses on a treadmill.

Between 1814 and 1819 Manhattan Island, New York had eight vessels operating from its shore some with 8 horses. The big disadvantage of this method of propulsion was horse dizziness.

Barnabas Langdon turntable patented design of 1819 enabled horses to walk tethered to one spot at the outside edge of the turntable which became the widest part of the single hull vessel. Forward of the horses were the paddles. The deck between was wide enough to take vehicles.

While documents and photos of these vessels existed it was the finding of the Burlington Bay Horse ferry [9] at the bottom of Lake Champlain in 1992 Figure 6] which stimulated research

The third method using horses was the treadmill which was developed in the late 1830. Tread mills are similar to short [continuous] conveyor belts found in exercise centres, an end roller axis being the power take off. They were normally positioned near the stern, sometimes in double formation.

The last known horse ferry remained in service until the late 1920 on the Tennessee River [15]

4.3 LAYOUT

There seems to have been three basic designs:
1 this required two hulls to accommodate the whims which were on the centre line. The deck width was dictated by the whim diameter. The paddle was between the hulls.
2 a single hull with a turntable just below he deck with a diameter great than the hull. The side paddles were forward of the turntable.
3 a single hull with treadmill [s] aft and paddles forward of amidships.

The overall dimension of these vessels can be assessed by consideration of machinery and horses requirements

The horse requirements vary from one to eight horses

1] The diameter of the whim and say four horses must have been of the order of 5m with access outboard of 1m each side say 7m. Only the smallest vehicles could have been carried as the side access was small.

2] The Burlington Bay vessel was a two horse vessel with dimension of

LBP	18.4m	(60ft 4in)
L deck	19.0m	(62ft 5in)
B mld	4.6m	(15 ft 2in
B at Deck	7.2m	(23ft 8in)
D mld est	1.65m	(6ft 0in)
T est	0.6m	(2ft 0in)

This vessel had ramps at the ends for vehicles and side enclosure for paddles and horses.

The transmission mechanism consists of a circular turn table set just below the deck, there being cut outs either side of the deck for the horses hooves. At one end of the cut out is a tether post for the horse to pull against

The turntable engages with a spur gear which then transmits power to a bevel gear and the paddles shaft. Paddle direction is reversible by a gear shift.

3} The horse tread mill would have been roughly the length and width of the horse say 2.5 x 1.9m.

The tread mill[s] appears to be in an aft side position which would have permitted vehicles to load at the ends. Some small vessel designs show double horse side by side treadmills.

4.4 DESIGN ASPECTS

Earlier vessels would have been constructed in wood with metal strips for strengthening

Most would have had double ends with short ramps, berthing to a quay the side paddles precluding side loading

Manning would consist of a steersman/master located at the stern for the second and third type. The earliest type had a forward and an aft rudder situated between the hulls so might have had two positon to control the vessel as the whim blocked any vision from the stern.
Another man would be required for the care of horses and subsequent loading and mooring. The drawings show a minimal outfitted vessel. It is said they were cheap to operate

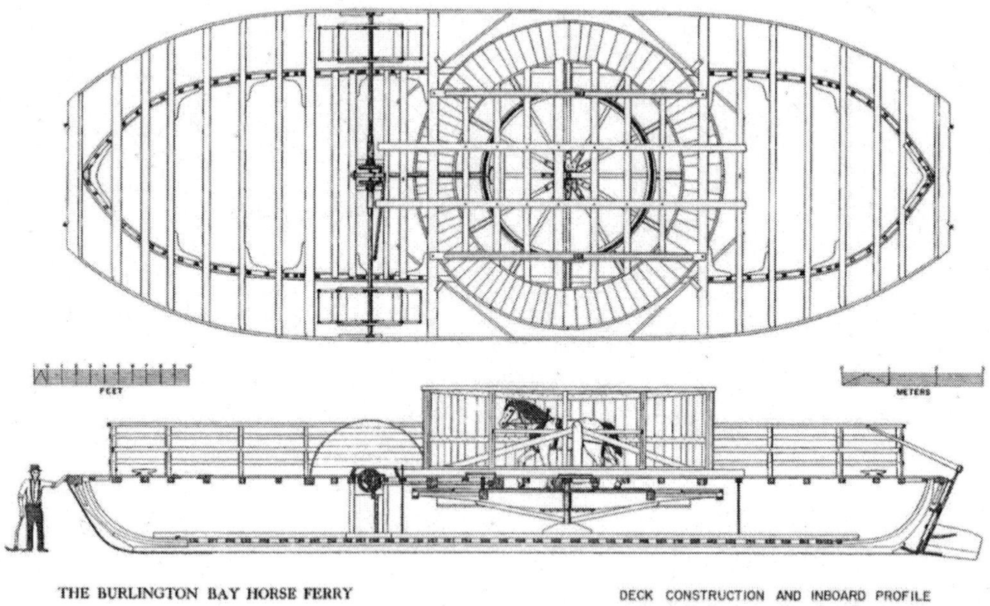

THE BURLINGTON BAY HORSE FERRY

DECK CONSTRUCTION AND INBOARD PROFILE

4.5 MODERN EXAMPLES

This section is included to be consistent with subsequent vessel types. There is however no existing horse propelled ferries in North America but there is half size Model on display at the BBHFW Museum.

It is interesting to contemplate building a full size vessel replica and how it might comply with modern regulation for Vehicle and Passengers vessels

5. PENDULUM FERRIES

5.1 CONCEPT

Pendulum Ferries are ferries that rotate from one side of a river to the other and back, being anchored half way between the banks. They can be considered to be one of two types, twin hulls (TH) and single (SH) with the following features.

Type TW
- the vessel is deployed longitudinally to the river and has side loading arrangements.
- it is anchored by the bows.
- it has large rudders..

Type SH
- the vessel is deployed transversely to the river and has end loading.
- it is anchored amidships with two heading lines for directional heading controlled by winches.

In Germany they were know originally as Fliegende [flying] Faehren but nowadays are known as Gier Faehre, gier being translated Yaw, and in the Netherlands Gierpont

Reference [18] says the first Fliegende Faehre on the Rhein was operating at Nijmegen in 1657 and was thought to be designed by Hendril Henck, followed by Mannheim 1669.

Figure 7 Mannheim Pendulum Ferry.

There is model of the Mannheim ferry in Stadtgeschichlichen Museum, Düsseldorf and a famous Kupferstich also in [18] which also shows the upstream anchor rope support boats.

The vessel shown above has twin hulls and a wide deck suitable for vehicles with side loading arrangements.

The bow and stern had rounded ends. At the stern are two long rudders with extra long tiller arms possibly connected together. There is an arch opening in the aft deck railings.

Forward are two masts [on other diagrams for this type of vessel there is a cross beam] with stays to the stern. The anchor cable is connected to the top of either mast depending on direction of travel hence the rigging which could be for moving the towing point.

The anchor rope is upstream supported by four boats with the final section dipping below water to the anchor[s].The anchor line is above the water level for most of its length presumably for preservation. Half one side of the river was blocked by the anchor rope at any one time.

To tack across a river the steersman would have to raise the short side ramps, disconnect from shore, move the transverse block on the cross beam of the masts to the opposite side and then put the rudders over.

Note the deck is completely open and only suitable for a few vehicles. Other vessel of this era had aft deck houses which were very ornate.

The present day Rathan ferry on the R. Elbe is probably the only existing ferry which has similar features to the type TW.

5.2 MOORINGS & RUDDERS

The mooring arrangement for the earliest vessel type TW is described above. The length of anchor rope was arranged to line up with the landing quay. The anchor rope length is of the order of the width of the River at that crossing point. Rudders have to have long length to obtain sufficient area as depths would have been restricted.

Type SH has become the basis for most modern ferries and is larger version of that shown in Figure 1. See also and Figure 9. The anchor line is connected to amidships and adjusted in length to coincide with slipways.

Rudders are not required as the hull is unsymmetrical to the river flow. In operation the end control wires orientate the hull to generate the transverse force. These are adjusted by winches.

5.3 LAYOUT

Timpe [19] gave carrying capacity of German Gier Faeren as 3 to 25 tons the hull length and breadth of the largest being about 20 x 5m.

Drawing tells us little about the hull construction of TH which initially would have been similar to horse propelled ferries.

The repeated grounding on a slipway of the SH type was probably one of the reasons for changing over to wrought iron and then steel construction.. A guide to Dutch steel inland waterways vessel construction can be found in [20] surprisingly it make no reference to Pendulum or to High Wire Vessel though it does discuss machinery for Ground Chain/wire vessels

5.4 DESIGN ASPECTS

End Ramps of SH Vessels are longer than TH because of the slipway and swim hull ends. The ramps are either counter balanced by suitable local weights both sides or by connecting the end ramps together through a high linkage and hand wheel so that while one ramp was level the other was in a raised position and vice versa. Finger ramps varied according to location.

SH vessel would have had controls [later wheelhouse] on the upstream side the down side being for an emergency anchor and boat.

Type TW needs a continuous watch system to operate so as not to interfere with other river users. Manning would be required to move the mast block from side to sides, raise the short ramps and mooring and there after steer the vessel.

5.5 MODERN EXAMPLES

The Location of these ferries on the Major European Rivers can be found in WESKA [21] for all Lorries [W] and Passengers [P] ferries explicitly stating whether they are gier or high wire or cable types with the exact position in km

The European Continent rivers and canals have kilometre markers erected according to the distance from the nominal source or State Border. e.g. Rotterdam by coincidence is at marker 1000 km from Konstanz. This makes the identification of ferry location and others features easy for all river users

5.5(a) Germany

The Deutsches Schiffartsmuseum, Bremerhaven (DSM) has a list [22] of all Ferries in Germany according to type being a successor to [20]

There are two rivers in East Germany which still have SH type ferries, the Elbe and the Saale.

River Elbe is the best place to see these in Germany which has 14 of which only 3 are self-propelled.

However there is a type TH passenger Ferry on the Elbe at Rathan in the Saechsische Schweiz which can carry vehicles and is very similar in design to Figure 7. It is very unusual in that it is situated on a bend, the anchor point being on right bank and not in the river. When the vessel is loading on left bank the river is blocked!
The River Saale has 3 SH vessels, the ferry at the crossing nearest to the Elbe taking up half the river width when loading.

Many of these TH vessels existed on the Rhein [24] as we have seen but none exists today, probably because of the heavy traffic which would have been disrupted by closing off half the width of the river most of the time.

Figure 8 show the German River and Canal system.

5.5(b) Nederland

Addition information on Netherlands vessel types and locations can be found [23] Some 11 vessels of type SH all with motors can be found on
River Ijssel
River Leck,
Nederrijn
Pannerdens Kanal
River Maas

5.5(c) UK

There are no vehicle vessels of this type in the UK.

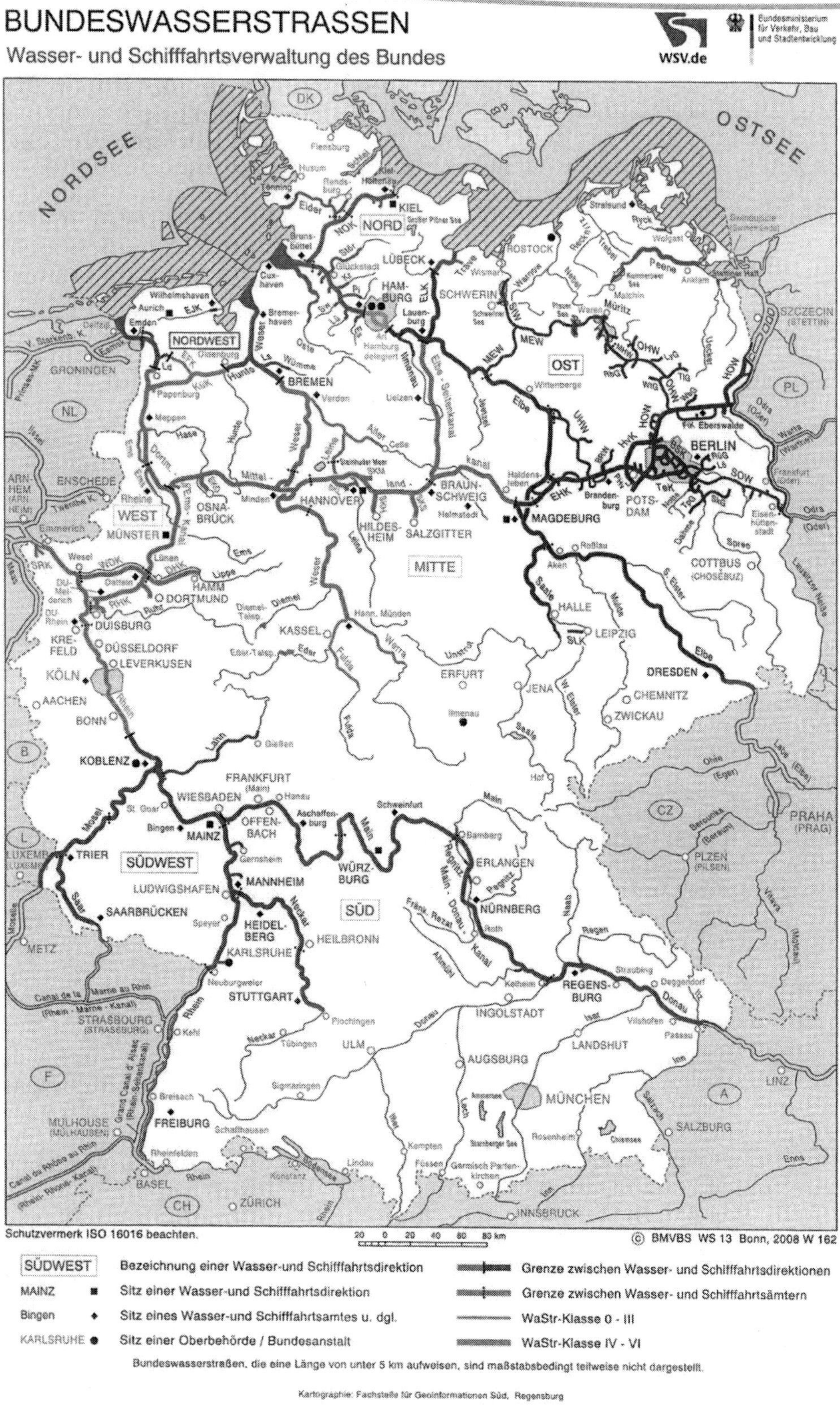
BUNDESWASSERSTRASSEN
Wasser- und Schifffahrtsverwaltung des Bundes
WSV.de
Bundesministerium für Verkehr, Bau und Stadtentwicklung
NORDSEE
OSTSEE
DK
NORD
KIEL
LÜBECK
HAMBURG
ROSTOCK
SCHWERIN
NORDWEST
BREMEN
Wilhelmshaven
Emden
GRONINGEN
NL
OST
BERLIN
POTSDAM
MAGDEBURG
BRAUNSCHWEIG
HANNOVER
SALZGITTER
HILDESHEIM
OSNABRÜCK
WEST
MÜNSTER
ENSCHEDE
MITTE
HALLE
LEIPZIG
DRESDEN
CHEMNITZ
ZWICKAU
COTTBUS
ERFURT
JENA
KASSEL
DORTMUND
HAMM
DUISBURG
DÜSSELDORF
LEVERKUSEN
KÖLN
AACHEN
BONN
B
KOBLENZ
FRANKFURT
WIESBADEN
MAINZ
OFFENBACH
TRIER
SÜDWEST
LUDWIGSHAFEN
MANNHEIM
SAARBRÜCKEN
HEIDELBERG
KARLSRUHE
HEILBRONN
STUTTGART
SÜD
WÜRZBURG
ERLANGEN
NÜRNBERG
REGENSBURG
INGOLSTADT
LANDSHUT
AUGSBURG
ULM
MÜNCHEN
SALZBURG
INNSBRUCK
FREIBURG
MULHOUSE
BASEL
ZÜRICH
CH
F
A
L
CZ
PL
PRAHA (PRAG)
PLZEN (PILSEN)
SZCZECIN (STETTIN)
LINZ
METZ
STRASBOURG
Elbe
Rhein
Main
Donau
Mosel
Neckar
Weser
Saale
Oder
Schutzvermerk ISO 16016 beachten.
20 0 20 40 60 80 km
© BMVBS WS 13 Bonn, 2008 W 162
SÜDWEST Bezeichnung einer Wasser-und Schifffahrtsdirektion
MAINZ ■ Sitz einer Wasser-und Schifffahrtsdirektion
Bingen ◆ Sitz eines Wasser-und Schifffahrtsamtes u. dgl.
KARLSRUHE ● Sitz einer Oberbehörde / Bundesanstalt
Grenze zwischen Wasser- und Schifffahrtsdirektionen
Grenze zwischen Wasser- und Schifffahrtsämtern
WaStr-Klasse 0 - III
WaStr-Klasse IV - VI
Bundeswasserstraßen, die eine Länge von unter 5 km aufweisen, sind maßstabsbedingt teilweise nicht dargestellt.
Kartographie: Fachstelle für Geoinformationen Süd, Regensburg

6. HIGH WIRE FERRIES

6.1 CONCEPT

The first High Wire ferries function the same way as the Pendulum Ferries but their anchor point was at the pulley on the transvers wire and not midstream.
In Canada these are called reaction ferries.

High Wire Ferries can also be considered to be one of two types, twin hulls (TH) and single hull (SH) and with the following features:

Type TW
- the vessel is deployed longitudinally to the river and has side loading arrangements
- it is connected to a transverse wire by the bows and not anchored
- it has large rudders

Type SH
- the vessel is deployed transversely to the river and has end long loading ramps
- it has two heading lines for directional control

The SH can be operated with two equal length heading lines ground chain ferries which is discussed in the next chapter.

It is believed that this development was primarily due to the increasing use of Chain Boats [25] in mid-19th century. These were self-propelled steam boats hauling in chain that had been laid on the river bed in certain European rivers and latterly in American. The Chain would have been inconsistent with the Gier Anchors so they had to be separated on each stretch that chains were laid. [26] This would have affected principally the rivers Elbe, Saale, Main, Neckar and Danube in both Germany and Austria. Chain boats disappeared in the early 20th century with the development of propellers.

The invention of reliable steel wire in the early 1830s by a German mining Engineer permitted greater spans at any height without restricting other river users. Where there are none or few other users then the transvers wire could be at about deck level and be disconnected. But this was inconvenient. Such a system may have worked with rope previously for short crossings.

6.2 HIGH WIRES

The height of the wire above the water surface depends on the height of other vessels and nowadays more than the CEMT waterway requirements for a Class V waterway and prevailing bridge clearances whichever is applicable. Where other users are involved the transverse wire could be disconnected temporarily.

6.3 LAYOUT

There are TW type ferries still on the Austrian, Danube. They are similar to the early Pendulum ferries in chapter 5 except they have a raised middle deck and no forward mast for connection to the single high wire. The control bridge is aft of the deck space. This part of the Danube was developed for hydroelectric power from 1938 and is complete upstream of Vienna where these vessel operate.

Figure 9 [26] shows a middle Weser powered SH type. It has a wheelhouse on the upstream side with two directly connected wires.

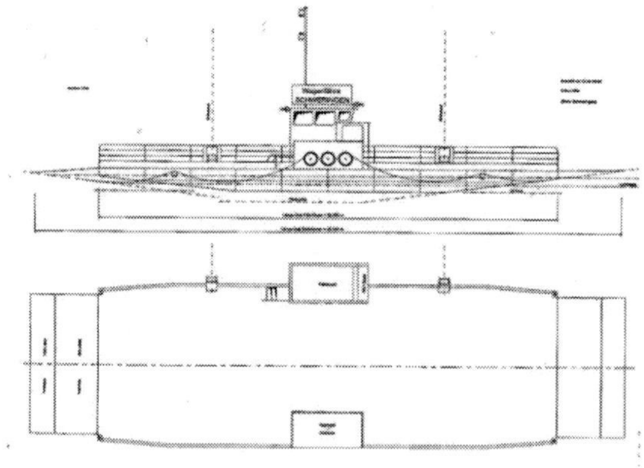

Figure 9 Proposed HW ferry for Schweringen

The vessel has hauling machinery on the downstream side. This consists of a three capstan heads. The hauling wire is supported along the side just clearing thee ramp.

6.4 DESIGN ASPECTS

Because of the height of the transverse wire there are no anchor boats obstructions as with pendulum ferries. Previous SH type comment from chapter 5 would apply. An engineer would not be required for this vessel.

6.5 MODERN EXAMPLES

Examples can be found in the following countries

6.5(a) Germany

All are SH type.

The Saale has the most high wire pendulum ferries.

The Danube has 3 of which the smallest can be found at Weltenburg.

The Weltenburg vessel is unusual in that the vessel axis is parallel to river.and is connected to the forward corner of the side to which it is traveling. This means it needs to be set before every journey. It is a one man operation and can take one car. A large Oar serves as rudder.

The Saale, Main and Neckar each has 2 wire guide vessels

6.5 (b) Netherlands

There are no High Wire vessels of this type in the Netherlands though there is highbred with anchor on R Mass between Southern Netherlands & Belgium

6.4 (c) UK

There are no vehicle vessels of this type in the UK

6.5(d) Austria

The Austrian Danube has three pendulum type TH ferries which are briefly described in 6.3

6.4(e) N America.

There are Reaction Ferries [HW pendulum] on West Coast Rivers.

7. GROUND CHAIN FERRIES

Rope was displaced by stud link chain first and then by steel wire. Chains had the big advantage that they would sink and lay on the river bed and thus permit others user unrestricted navigation.

The original steam chain ferry was designed by James Meadow Rendel a Civil Engineer who subsequently set up Rendel Palmer and Tritton a famous firm of Consulting Engineers for whom I subsequently worked.

7.1 CONCEPT

Rendels original Floating Bridge, a term he invented, was built in 1831for the river Dart crossing [27] It comprised twin hull 40 x 30 ft. overall with a 4hp stem engine mounted between the hulls and two chains. Each side had a overhead counter balanced ramp. Rendel was responsible for most of the early UK Chain Ferries.

A continuous chain drive man hauled from the shore had been considered for the Dart! This actual vessel subsequently had its machinery replaced by a horse whim [28]!

Modern designs in general prefer either two chains being sited inboard of any side tanks or with pairs on the outside.

Figure 10 shows the arrangement for the new Cowes chain Ferry [29]. It features single ramps at each end with passenger accommodation on one side.

7.2 CHAIN

Stud chain was patented in 1819. The coming of steam machinery at this time permitted hauling power to increase steeply and eventually larger vessels.

Rendel floating bridges had two chains. A later European design had one and a recent Canadian design had three, two being for guidance.

Chains could be manufactured for any length as was seen by Chain Boats [26].The longest cross river ferry chain in the world is thought to be 1.9km at Buckley Bay, Vancouver Island. Canada.

Chains are however noisy in operation. In time they have been displaced by wire and positioned above the deck.

7.3 LAYOUT

Large ferries often have a reverse sheer deck sloping down to the end ramps. The hull is subdivided throughout and fitted with side tanks and voids spaces A large chain ferry would have hull dimensions of 73 by 20m and a draft of 1.6m.

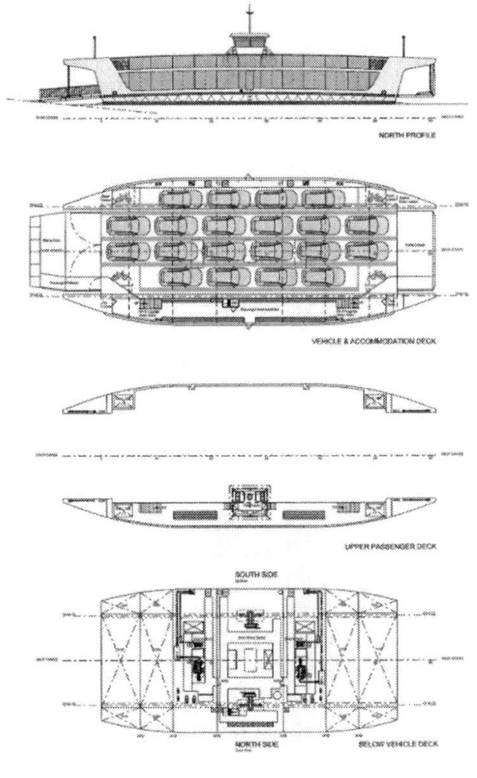

Figure 10 Cowes Chain Ferry

7.4 DESIGN ASPECTS

Ramps nowadays are hydraulically operated being suspended from horns.

The most interesting part of the machinery is the chain gypsie which is at the centre of the length the top which is close to the deck and is powered by hydraulic orf electric motors.

Wire guided ferries can have machinery on the deck. The present Dartmouth ferry is wire guided and propelled by paddles.

In the UK the responsibility for the design and operation of for these vessels comes under the Health and Safety Executive but by an Agency agreement delegates functions to the Maritime and Coast Guard Agency. This body has a Code of Practice [30] for the construction, machinery, equipment, stability and operation of Chain/wire ferries, carrying passengers and vehicles.

7.5 MODERN EXAMPLES

Examples can be found in the following countries

7.5(a) Germany

These vessels can be found on
Markiscke Gewassser
Rhein
Ruegen

7.5 (b) Nederlands

No high wire vessel exist in the Netherland though there is a hybrid on the Belgium Maas

Some 11 Vessels of type SH can be found on
River Ijssel
River Leck,
Neder rijn
Panner dens Kanal

All 24 have been motorised.

7.5(c) UK

The UK hass six chain/wire ferries four have river locations, one on a Lake and another on an Inland river. Plymouth has three ferries in parallel. The MCGA has Code of Practice for these vessels

7.5 (d) N America

Ground wire ferries can be found on the East Coast of the USA and the East and West Coasts of Canada

8. CONCLUSIONS

1 There is a lot of useful information on ships and other vessels that can be obtained from old Paintings and Photographs but no easy way of accessing it for specific types.

2 The Royal horse propelled vessels of 1680/2 may be the earliest vessels of their type in Northern Europe. A painting or some Royal record may provide more information to confirm this.

3 Further work is need to be done to substantiate details of the 1827 Amsterdam Ferries as it appears they are unique with a horse engine room below deck.

4 Pendulum ferries continue to provide a reliable cross river service though most now have diesel assist.

5 Horse Propelled vessels, Paddle Boats and Chain Boats have been consigned to history but made important contribution to the transport system at the time.

6 Chain, wire and stem propulsion caused major changes in ferry design at their introduction.

9. ACKNOWLEDGEMENTS

With thanks to C Gardner and R Wieffer with computer and translation assistance and :

- National Maritime Museum, Greenwich
- National Maritime Museum, Cornwall
- Binnenschiff Museum, Duisburg - Rurort
- Deutschland Maritime Museum, Bremerhafen
- Het Scheepvaart Museun, Amsterdam
- Maritime Historical Museum, Rotterdam

10. REFERENCES

1. Moyse A D, The Operation of Coasters into the European Inland waterways, Hydraulics Research Station, Wallingford Oxn 1978.

2. Faehre von Worms um 1645, Aus einem **Kupfersrich** von Mattjaeus Merian D. A Cover picture on: Faeher Neckarhausen Geschichte und Geschichten Gemeinde Edingen-Neckarhausen

3. In de strengen 19.5.1977, Paard en Scheepvaart, Part 1 & 2

4. Stursburg U, Innovation auf dem Rhein, Das End der Treidelschiffahrt, Klartext verlag, Essen, 2015, ISBN 978-3-8375-1350-9

5. Rontgen J F, Het Schip Utopia, Bussum, 1983

6. Savery T, Navigation Improved: London, James Moxon, 1698

7. Scheepshistoric 10, S & History,Lanasta

8. Corlett E C B,The Iron Ship, Moonraker Press 1978, SBN 239.00112.5

9. Crisman K J & Complain A B, The Burlington Bay Horse Ferry Wreck, May 1993, Lake Champlain Maritime Museum, Vermont USA.

10. Singel et Al, History of Technology Vol 11, Oxford, 1979

11. Stuart R, Anecdotes of Steam engines and their Inventors & Improvers by , Wightman & Comp, London, 1829

12. Ons Amsterdam, Maandblad van de Gemeentelijke Commissie Heemkennis, Jaagang Ni11, Noverber 1967

13. Dehe J, Een slaafsch en ongezaond bedrijf, De geschiedenis van het openbar vervoer in Waterland , 1630 -1880, Hilversun, Verloren, 2005

14. 'Experiment'(horse-powered boat), Wikipedia, the free encyclopedia, 2016

15. Crisman K J, When Horse walked on water: horse-powered ferries in nineteenth-century Amerika Washington: Smithsonian Institute Press, 1998, ISBN 1560988436

16. Dunnigan C, Horse-powered Ferries: Discussion Keys Waterway Transportation_ (http://m.mackinacislandnews.com/news/2012-08-04/columnists/Horse_Tales.html), Horse Tales, Mackinac Island Town Crier, August 4, 2012)

17. Baxter R J & Adams A G, Railroad Ferries of the Hudson, Lind Publications, ISBN 0-910389-01-2

18. Kludas A & Schnall U, Deutsches Schiffahrtarchiv 6 – 1983, Stalling Verlag Gmbh, Oldenburg 1983

19. Werner T von, Die Gierseilfaehren in Deutschland, Hannover 2001

20. Roorda A Ir, Small seagoing craft and vessels for inland navigation, The Technical Publishing Company H Stam, Haarlem, Holland 1957

21. Westeuropaischer Schifffahrts und Hafenkalender [WESKA] 2015, Binnenschiffahrts-Verlag Gmb, Duisburg-Ruhrort, Deutschland

22. Erlaeulrernubgen zur DSM-Faehrdatei ab2015, Deutsches Schiffahrtsmuseum, Bremerehaven

23. Vrienden van de Voetveren, Dutch List

24. Boch W, Die Geschicht der Rheinschaffart

25. Chain Boat Navigation, Wikipedia, the free encyclopaedia, 2016

26. Kroeger L, Crossing the River: Ferries as part of the maritime landscape for the Main river, Germany, Internationalsymposium on boat and ship archseology, ISBSA 13, Amsterdam

26. Brandt F-W, Fahre Schweringen, die letzte schwimmende Brucke der MittelweserAarbeitgemeinschaft Binnenfahre in Deutschland Schriftreihe Band 2, Deutsches Schiffahrtsmuseum, Bremerehaven, 1999

27. Lane M R, The Rendel Connection, Quiller Press, ISBN 1 870948 01 7

28. Kittridge A, Rendels Floating Bridges, Twelvehead Press, Truro, 2008, ISBN 978 0 906294 67 0

29. Ship and Boat International, RINA, July - Aug 2016

30. MCA Code of Practice for the construction, machinery, equipment, stability and operation of Chain/wire ferries, carrying passengers and vehicles.

11. AUTHORS BIOGRAPHY

The author retired in 2002 having been a Principal Inspector in the Offshore Safety Division of the Health and Safety Executive, and the Offshore Certification Bureau.

He worked as a Naval Architect in various Consultancys including Rendel, Palmer and Tritton where he became involved with Port Engineering

Most of his earlier career was spent in the design of naval auxiliary vessel, Supply, Tugs & Trawlers, and as a Ship Surveyor with the Board of Trade..

He designed the Hammond Innes the largest UK wet fish stern trawler while Naval Architect at C D Holmes, Hull.

THE SHIPSHAPE HERITAGE TRAINING PARTNERSHIP (SHTP) PROJECT

H Cunliffe, National Historic Ships UK, United Kingdom

SUMMARY

This paper describes the Shipshape Heritage Training Partnership (SHTP) project, grant aided by the Heritage Lottery Fund (HLF) and set up in 2014 to develop traditional skills and techniques in young people, focusing on the conservation, handling and maintenance of historic vessels. The project was led by National Historic Ships UK (NHS-UK) which maintains the National Register of Historic Vessels listing over 1,200 craft in need of ongoing conservation. The SHTP project provided ten 12-month training placements at five partner sites across the UK, offering on-board specialist training to ensure the significance of these craft and the ability to operate them safely and effectively was kept alive. This paper reviews the SHTP project, highlights the traditional skills at risk of being lost, and features case studies from the trainees and partner bodies involved, as well as explaining the outcomes which form the legacy of the scheme.

1. INTRODUCTION

NHS-UK is a government funded, independent organisation set up in 2006 to provide objective advice to UK governments and local authorities, funding bodies, and the historic ships sector on all matters relating to historic vessels in the UK. A principal function is to maintain: the National Register of Historic Vessels (NRHV), currently listing over 1,300 craft meeting certain basic criteria; the National Historic Fleet, formed of some 200 vessels of pre-eminent national significance; the National Archive of Historic Vessels (NAHV) for those vessels which have now been broken up or lost; the Overseas Watch List (OWL) for significant UK vessels based abroad, and the UK Replica List. In addition to holding and servicing these databases, NHS-UK is concerned with the infrastructure in place to support these vessels in the future, including the facilities to maintain them such as slipways and dry docks, the traditional seamanship techniques needed to operate them in accordance with their significance, and the conservation skills to care for and preserve their fabric.

To promote the regeneration of these facilities, skills and techniques, NHS-UK set up the Shipshape Network in 2010. This initiative provides an online framework at www.shipshapenetwork.org.uk for all those with an interest in ship conservation, is home to the National Directory of Skills & Services, offers a jobs vacancies page and promotes training schemes and other projects relating to historic vessels on the National Registers. Under the umbrella of this Network, NHS-UK put forward an application to the Heritage Lottery Fund's Skills for the Future programme in 2012, successfully securing a grant of £261,100 for its SHTP project. This project sought to promote the skills of operating, maintaining and conserving historic vessels through ten 12-month training placements.

2. THE SKILLS GAP

In the years leading up to this project, NHS-UK was increasingly approached by vessel owners and custodians who were struggling to find the skilled seamen they needed to operate their vessel in the manner for which it was built. This was due to the fact that those retaining the specialist knowledge to handle these vessels were the last generation using knowledge and techniques passed down from seamen who worked on craft by apprenticeship and which were gained through decades of experience. The current generation needs the same level of training and experience, to enable them to operate a boat safely and harmoniously, to 'feel' the vessel and its progress. This can only come from being consciously competent and learning how to use ropes, tackle and tools to work with the boat and the elements. The operational skills still exist, but are in danger of being lost as those with the knowledge age and modern technology grows, or being carried out ineffectively by users who do not have a full understanding of the techniques involved. These skills cover almost every aspect of vessel operation and range from daily disciplines (where the logic and purpose is being forgotten – such as the reason why we need to coil ropes clockwise) to large and complex operations where very few people now retain the skill, such as dipping the lug to tack a large lug-rigged boat.

There is also a similar difficulty in terms of traditional maintenance skills: every choice and every action on a historic vessel is performed as a result of generations maintaining and operating it for optimum performance. The skills to perform tasks are largely still extant, but the reasoning for making a particular choice is disappearing fast. Whether to profile caulk pitching 'paying up' or 'paying down' is a decision founded in the performance of the materials in different conditions and different waters – a wrong decision can significantly affect the life expectancy of deck timbers, a coat of plastic paint can cause serious deterioration, a repair with the wrong metal can, through chemical reaction, destroy an historic artefact. Before SHTP, private owners of historic vessels also had no mechanism for learning the techniques to maintain the craft themselves.

In 2007-8, NHS-UK carried out an Infrastructure Audit looking at skills shortages in the historic vessel sector. Of the 109 responses received, 54% had difficulty sourcing the traditional skills they needed. The survey revealed a keen interest amongst owners for training, particularly in the following areas: repair techniques, rigging, sail making, canvas work, engine repairs, historic vessel operation, vessel research and maritime conservation. A later survey of vessel owners in 2009 indicated the profile of those working in the sector was 45+, with 51 responses showing an aging workforce. Further research was carried out by NHS-UK in preparation for the SHTP project and a working group was set up covering vessels with gaff, lug and sprit rigs comprised of five industry experts with knowledge of both traditional seamanship and modern sailing. This group identified the skills specific to handling vessels with specialist rigs and these were tested holding two pilot training days on registered historic vessels. The key skills were documented into a written framework, which later formed the basis of the skills passport that the SHTP trainees' learning was based on.

None of the existing mainstream providers of training run a course that identifies the specific skills in handling, operating and maintaining an historic vessel - some courses cover some aspects of this and there are courses which teach sailing using modern vessels. Training organisations in the sector focus on sailing or boat maintenance, but do not meet the need to teach sailing and maintenance within the context of historic vessels which are so different in fabric, operation and original use. Prior to the SHTP project, training for the specific skills required to operate an historic vessel had not been collated or addressed in any existing training course and was not within the range of any accreditation or certification scheme, with the exception of the Bargemaster's Ticket for sprit rig training, which is run on behalf of the Maritime & Coastguard Agency by a joint panel comprised of representatives from the Association of Bargemen and the Sailing Barge Association.

3. PROJECT OVERVIEW & RECRUITMENT

To address the skills gap as set out above, NHS-UK put together an application to the Heritage Lottery Fund's Skills for the Future programme in 2013. This programme was set up to provide funding for heritage skills in the workplace. Supporting 93 projects and investing £47 million across the UK in total during 2010 and 2014, it was an ideal match for the type of training needed. The application was successful and £261,100 was awarded by HLF. This sum was match funded with a contribution of £3,800 from NHS-UK's own resources, as well as £2,000 from the International Guild of Knot Tyers towards the delivery of masterclasses in rope work, rigging and sail repairs. £8,000 was awarded by the PRISM (Preservation of Industrial and Scientific Materials) Fund towards the skills mapping exercise and a further £12,000 funding was secured from the PRISM Fund and put towards conservation work at each partner site involving the historic vessels within the project, for works which could be carried out or observed by the trainees.

The SHTP project was delivered by NHS-UK in partnership with four sailing organisations and their respective historic vessels: Trinity Sailing Foundation (Brixham trawler *Leader*), Excelsior Trust (Lowestoft smack *Excelsior*), Sea-Change Sailing Trust (Thames barge *Reminder*) and Dauntsey's School (pilot cutter *Jolie Brise*), with the Scottish Fisheries Museum (fifie herring drifter *Reaper*) acting as the fifth partner to offer conservation expertise and provide a lug-rigged craft. The five partner organisations differ in their core purpose, organisational structure and type of historic vessel, each offering a slightly different learning approach and outcome. They were chosen for this reason, to reflect sprit, lug and gaff rigs, cover a wide geographical area, provide different business models and for their previous working relationship with NHS-UK which enabled the project team to put the application together in the relatively tight timescale set by HLF.

The project was set up to run 12-month long traineeships for five trainees per annum over a two year period as set out in the SHTP Training Plan written during the development phase. The overall management of SHTP was by Hannah Cunliffe, Policy & Project Manager for NHS-UK who wrote the original Heritage Lottery Fund application and set up the programme. A part-time project co-ordinator was recruited to deliver day to day management. Both these positions reported to the NHS-UK Director and to a steering group which comprised a representative from each partner organisation. The project co-ordinator and staff at NHS-UK managed the project budget and reporting process to HLF. Under this budget, trainees were employed as full-time staff and were paid a monthly sum to cover their living expenses and accommodation. Travel expenses incurred as a result of the project programme were also met from this source, with an additional allocation made for trainee personal development and tools.

As the project covered the whole UK, with host sites ranging from Scotland to the West Country and involved relatively young trainees working largely in isolation, a mentor scheme was set up to help with pastoral care. Mentors were chosen from the NHS-UK Council and from representatives within the Shipshape Network regions and were assigned to their trainees. Whilst for the most part, the trainees did not lean heavily on their mentors, there were cases where the mentor reported back to central management aspects of the placement which the trainee was struggling with or where mentors accompanied trainees on external visits to give further background to the historic ships world.

Recruitment was carried out with support from Royal Museums Greenwich human resources team under a service level agreement. The strategy was to encourage applications from those people who were planning a future in working with historic vessels either as part of a sailing crew or in terms of the care of the vessel through conservation or museum work. Because there was an identified need to pass on knowledge and skills from older workers in the sector, trainees were sought between the age of 18 and 30 in order to train the next generation of crew. In particular, through targeted advertising, NHS-UK was looking to recruit applicants who were female, in this age group and from a range of backgrounds to address current gaps in the demographic profile. With this in mind, the training vacancies were advertised as widely as possible with coverage on the NHS-UK and partner websites, maritime, sail training and heritage websites, as well as through the regional job centre pages, specialist boatbuilding colleges, sailing clubs and sea cadets.

Interviews for the trainees were carried out on the Thames barge *Lady Daphne* in St Katharine's Dock over a two-day period with an interview panel comprised of the Director and Project Manager from NHS-UK, HR personnel and a representative from each partner organisation. Due to the time of year and the short timescale for recruitment it was not possible to organise a seamanship assessment as part of the recruitment process, but a short practical was held with each candidate on deck where they were asked simple questions about the vessel and the rig. 53 applications for five traineeships were received in year one. At the end of the project, nine out of ten trainees recruited met the age criteria of being between the ages of 18 and 30, and two of the trainees were female.

4. THE TRAINING PROGRAMME

Each trainee was allotted a host from amongst the partner organisations and remained with that organisation for at least six months of the placement. During this time, they were considered trainees of the organisation they were with and were able to explore areas they wanted to learn, given support from their supervisor on site, as well as opportunities to work with different audiences, taking part in charters, sailing races and on board activities. Throughout the season, all trainees went on two-week long rotations allowing them to experience at least two other partner vessels. In each rotation, trainees worked as one of the crew and were given opportunities to use similar skills but apply them to different kinds of vessels appreciating the differences between each vessel and why these exist. The rotations also gave trainees exposure to different historic environments, business models and an introduction on how to interpret heritage to a wide range of audiences, including visiting groups and charters.

All trainees also undertook a course in historic vessel maintenance run by the International Boatbuilding Training College (IBTC) in Lowestoft. This was split into two sections – covering essential skills at the start of the programme, as well as a general understanding and background to the vessels, whilst the second period was set at the end of the sailing season to build on knowledge gained through operational experience and provide new skills needed for the winter refit. The initial course covered the use and maintenance of hand and power tools, traditional tools such as caulking irons, basic joinery skills, traditional rope work, basic sail repairs, decision making in repairs and maintenance, learning the properties of different materials and repair techniques. Skills taught in phase two included: use of paints and varnishes, caulking seams, three strand splicing, running repairs, use of pitch, spar building and spar hardware, properties of an historic vessel, safety on vessels, living on board, general deck work, watch, navigation and meteorological skills, ship husbandry.

Evaluation showed that all trainees valued this element of the programme, seeing it as an opportunity to work together, explore skills in more depth and learn from experienced boat builders. Each trainee received a joint certificate from IBTC and NHS-UK recording their achievement. Trainees were also offered other qualifications for personal development and to help with career prospects, including: MCA Diesel Engine Maintenance course, Powerboat Level 2, VHF License, Personal Survival Techniques, RYA First Aid, STCW Elementary Fire Fighting, Personal Security and Social Responsibility. Masterclasses in sail repairs, knotting and wire splicing were also carried out, with experts including the sail maker Steve Hall and rigger Nigel Gray. These also provided an opportunity to upskill staff from the partner vessels and train staff from IBTC in specialist techniques to enable them to carry out this type of training in-house in future.

All trainees undertook a month-long placement at the Scottish Fisheries Museum to give them an understanding of conservation and interpretation, as well as the significance of historic vessels. This included specialist maintenance tasks using parts of the collection and displays, conservation and interpretation principles, sector basics, introduction to museum and operational vessel models, fundraising, business planning, volunteer management, presentation skills, field trips to a range of operational and static vessels, ships owned by trusts, volunteer run projects, as well as successful commercial enterprises and skiff rowing.

The SHTP project was promoted via the Shipshape Network website with each partner allocated a project page in the relevant region and trainees uploading a monthly diary of their experiences. Social media proved a useful way of interacting with the trainees whilst they were at sea and an end-of-programme Networking Day brought the project to the attention of the sector, with 60 representatives attending. In the second year, a short film was produced entitled 'Learning the Ropes' which summarised the key objectives of the scheme and included interviews with the trainees. This was filmed during a group sail which brought all the second year trainees together with the project co-ordinator on board Thames barge *Reminder*. The film was published on You Tube and proved an excellent promotional tool.

5. CASE STUDIES

Hugh Langworthy joined the programme as a year one trainee with previous experience of working alongside shipwright Will Stirling and some knowledge of the skills needed to conserve historic vessels. He was interested in learning more about the operation of vessels and was placed with Trinity Sailing Foundation. Reflecting on his time as a trainee, Hugh said: *'When I saw this opportunity I jumped at the chance. I wanted to learn something new - I wanted more experience of sailing and SHTP helped me to learn what it was I wanted to do. I enjoyed the variety - there has not been anything like this before. Rotations were a great idea to have different experiences on each vessel. What I learnt were core skills although these change with different skippers. Most boat building courses were more than I could afford and could not offer the breadth of training on offer through the Shipshape project'*. Hugh now works as an apprentice rigger for Tommi Nielsen & Co in Gloucester on a two year term and summarises: *'I have learnt a lot of transferable skills I can use in the sector. I am now doing an apprenticeship to be a rigger - I see all parts of the work in the yard - there is variety in my work. They were aware of the course when I applied which showed them what a range of experience I had'*.

Catherine Holt, another year one trainee, said: '*I was looking for maritime skills and the opportunity to get hands on training - intense experience quickly. I have a background in archaeology and a Museum Studies MA and wanted to stand out as a candidate at interview. The training helped give me authority when talking about sail-making and specific tasks. Volunteer management and my personal project gave me a level of new responsibility to take on more in my job as Project Officer at Chatham Historic Dockyard. I learnt about the world of operational vessels and certain qualifications and skills expected which is very useful to marine archaeology and maritime collections. It was a great project and I thought the process was so well considered. I am very proud to have been part of it.'* Catherine was the first trainee to gain employment following the programme and worked as Project Officer for the Chatham Historic Dockyard which has since been followed by a promotion to Assistant Curator for the Princess of Wales's Royal Regiment & Queen's Regiment Museum.

The project partners also identified significant benefits from being involved in SHTP and when interviewed later, many described being able to connect better across the sector, learning more about how to deliver training and revisiting how they approach their own work. Richard Titchener, the Executive Officer and Primary Skipper from partner organisation Sea-Change Sailing Trust commented that: *'There is nothing like this course seen from an employer's point of view. This was a bold experiment and the first time it has been attempted - to have no dropout rate and almost all trainees in the sector at the end is an enormous tribute to the planning and running of the project. The sector now has a pool of knowledgeable people in the next generation who can actively network as individuals and expand their connections with each other.'*

For NHS-UK, as lead partner in the scheme, the SHTP project was a way to bring together five organisations that might not otherwise have had the capacity to organise and administer a training project of this kind or scale. It had enormous value for NHS-UK in terms of raised profile, developing key relationships with partners and other leading bodies, the skills mapping exercise which was an integral part of the scheme and in allowing NHS-UK to take the first steps in addressing the identified skills gap. NHS-UK also benefited from the experience of managing a project on this scale for the first term in terms of learning and evolving best practise.

6. PROJECT OUTCOMES

A full evaluation was carried out at the end of the project, based on data gathered throughout, with support from an external consultant recruited for the purpose. This allowed the project team and partners to carefully review what had been achieved, lessons learned and the key outcomes from the scheme. The formal evaluation report was published on the NHS-UK website and a summary leaflet was also produced for circulation to the sector.

It was concluded that the programme had been a successful example of a Skills for the Future project where heritage skills had been transferred to younger entrants in the historic vessel sector who are now using those skills in their career and will be able to continue promoting and passing on their knowledge within the sector. In addition, they have a strong understanding of the background to the life of historic vessels, their fabric, significance and their existing and potential uses today. Several trainees now work in conservation of maritime heritage having operated vessels or work in roles developing craftsmanship in skills associated with historic vessels.

All ten trainees successfully completed the SHTP programme learning the core skills of conserving, maintaining and operating historic vessels. All have completed boat maintenance modules, a conservation placement and placements working on board historic vessels. All have Skills Passports and five completed detailed personal projects. Nine trainees are now in paid and voluntary employment in the maritime heritage sector, trained in operating historic vessels with contacts and experience of working for five leading maritime organisations in the UK

New learning units in Historic Vessel Maintenance have been developed and can be delivered as part of a future course at the International Boatbuilding Training College, Lowestoft. A skills mapping exercise was carried out over the duration of the project to research and record all the skills being learnt by the trainees. This was undertaken by David Tournay of Evolution Partnership and a series of competency units was developed for future assessment of trainees and to improve the standard of training across the sector. These have been clearly laid out as a training workbook by an external design firm and are available for organisations across the sector to download from the NHS-UK website free of charge.

In response to feedback from the sector on the difficulties of recruiting skilled crew, a national crew swap online system was launched via the Shipshape Network website to create a crew bank of experienced individuals, match vacancies and enable an exchange of skills between different vessel types. An alumni association - MAST (Members and Shipshape Trainees Association) – was also created to sustain sector networking amongst skilled practitioners on historic vessels and keep NHS-UK in contact with the trainees and individuals associated with this and other similar projects. The first MAST Association event was held in Greenwich in November 2016, bringing together participants from the SHTP project with others working in the sector and offering them further skills development through workshops run on site by instructors from the International Boatbuilding Training College, Lowestoft.

7. CONCLUSIONS

In conclusion, it is clear that this project, as the first of its kind, has energised the sector to continue addressing the skills gap, and there is a marked enthusiasm from its partners to carry on working together. The project also identified the crossovers and connections between those operating historic vessels and those conserving historic vessels, as well as highlighting a further skills gap where maritime curators often lack practical experience in operating and maintaining historic vessels which they can then apply to conservation theory. As a result of this, a new application has been made to the Skills for the Future programme for a second phase of the scheme. This would see a further 16 trainees recruited across the UK and placed with existing partners, plus an additional three new partners – Royal Museums Greenwich (*Cutty Sark*), Pioneer Sailing Trust (smack *Pioneer*) and HMS Warrior 1860. This scheme offers two training routes with a focus on either traditional seamanship or maritime curatorial skills, whilst still retaining a strong crossover between the two, with both sets of trainees exposed to the skill-sets surrounding operational and static vessels.

The outcome of this bid will be announced in February 2017 and, until this time, NHS-UK continues to keep in close contact with its partners and support further ad-hoc training in the sector through the documentation which SHTP produced. This paper has summarised the SHTP training programme and its achievements as a pioneering scheme within the historic vessel sector. Whether the application to HLF is successful or not, NHS-UK hopes that the SHTP project has caused a step-change in the sector, with partner organisations recognising the value of what can be achieved by working more closely together and the need to highlight the decline in traditional skills. In the meantime, our ten trainees continue to work within the sector with heightened awareness of maritime heritage and the traditional skills associated with the conservation of historic vessels. The outcomes from the SHPT project have spread beyond the five regions where the learning took place and will continue to ripple through the historic vessel world in the years to come.

8. REFERENCES

1. National Historic Ships UK, 'Shipshape Heritage Training Partnership: Impact of the trainee programme on the historic vessel sector', *www.nationalhistoricships.org.uk*, 2016.

2. National Historic Ships, 'Conserving Historic Vessels', *Understanding Historic Vessels Vol. 3*, 2010.

3. Kentley, E, Stephens, S & Heighton, M, 'Recording & Deconstructing Historic Vessels', *Understanding Historic Vessels Vols 1&2*, 2007.

9. AUTHOR'S BIBLIOGRAPHY

Hannah Cunliffe holds the current position of Policy and Project Manager for National Historic Ships UK. She works to raise the profile of the UK's maritime heritage by: undertaking case work relating to vessels on the National Registers; seeking protection for the National Historic Fleet; managing the Shipshape Network and associated skills projects; preparing funding bids and providing expert advice to the Heritage Lottery Fund and other grant-giving bodies as required. Before accepting the role in April 2009, Hannah completed a Masters in Maritime History at the University of Greenwich and worked as a freelance maritime researcher and consultant for ten years. She spent her childhood sailing on a Bristol Channel pilot cutter.

RESTORATION HISTORIC/CLASSIC VESSELS: ENHANCING CAREERS FOR A BRIGHTER TOMORROW

R. Neish, Maritime Training Academy, UK

SUMMARY

The dream of many with an interest in boats/ships and the sea is to help to restore a fine old lady of the sea back to her original state and see the vessel as she was originally intended. We believe that many more people could be enticed into this exiting and rewarding field of maritime realisation, if they felt that the knowledge to do so, and the guidance to do so was provided in a clear and concise way as seen by men and women who have worked in the industry of building ships from the shop floor to the board room.

Having worked with many such people in the past I realised how frustrated many have become as there is little help to bring that dream to fruition and allow willing individuals to take the steps required to fulfil their own vision.

1. INTRODUCTION

A course produced with clear and easy to follow guidelines, and also help them towards a fuller understanding of the whole process.

To overcome this I set out to develop a course that will further the knowledge of Ship/Boat restoration, conservation and on what type of ships/boats to try and restore.

It is not intended to be the ultimate in ship/boat restoration and conservation, rather an overall guide to what steps you should be able to take to achieve your quest of saving that old ship/ boat from the scrapheap.

Most people and particularly those who tend to volunteer for good work tend to have busy lives. Therefore I divided the course into 10 modules so that people can study at their own pace. At the end of each module there is an assessment exercise to test understanding and if desired students can sit a formal examination at the end. Learning is guided throughout by individuals who share the same passion. All presented on-line with guidance on how you can participate in this field as a craftsman or manager or volunteer.

2. WHO IS THE COURSE SUITABLE FOR

Anyone with an interest in old ships, sail, steam or diesel, Experienced or amateur alike, will help with maritime history/research.

It may be suitable for anyone who is undertaking part in any kind of Marine Engineering or Apprenticeship and wishes to further their own knowledge on this wide subject. We aim to show how inclusive this field can be and not exclusive as some may perceive.

3. SETTING A COURSE

Restoration of an existing vessel not only preserves the vessel either as a static or waterborne exhibit even better if she can be returned to working order and go once more to sea as originally intended, the process also offers up so many other opportunities for historians or potential future historians to academic studies on through to providing some much needed work for skilled men down to the realisation of these skilled men being able to pass on the skills to younger people who may be floundering in the chase to find out just what they may like to do with there own futures, after all if you learn a trade you can always go back to it in the future, but that future has to be there for them.

Training and the passing on of knowledge is no use if there is no final outcome where this knowledge or skill can be used, if this was the case then you would as well be good staying at school for the whole of your life, it has no real tangible end product.

I myself am and have been personally involved in the attempts to restore the only remaining ice-breaking tug in this United Kingdom and she is also listed on the Historic Ships register, she is of riveted construction and to restore her will require the expertise of people with the skill of riveting, and it would be hoped that this skill or at least a basic knowledge of the skill could be passed on to interested parties, but then we are faced with the challenge of not only finding the required skills but once passed on where are the future projects to enable such skills to continue, or are we forever to be in the situation of seeking out a very few individuals to do such work, pass on some of the skills and then such skills are forgotten about due to lack of use.

If such a scenario is continued then perhaps we should not expect any other result than to continue to repeat the mistakes of the past rather than learn and progress from the past.

We are fortunate that some of the old skills do still exist and there are one or two small shipyards left around the country such as the Albion Dockyard in Bristol which reconstructed the Medway Queen.

Sadly the Albion Dockyard has also now closed down after its owner Mr David Abels decided to give up the lease only in September of this year, so yet more skills and experience is lost.

Figure: 1 SCOT II the only ice-breaking tug on the historic ships register in the U.K. (Photograph from authors own collection)

Such is the broad range of training possibilities associated with the restoration of old vessels from the project management aspects down to the lesser known old time skills fast disappearing that many of these skills can be used as a stepping stone into other parts of the maritime world which have already been identified as having a serious shortage of skills, with a recent report released by British Marine into the super-yacht industry, one of the real growth aspects of a once proud industry, showing a real concern at the future lack of skills that could have a serious impact on this industry in continuing to grow and I show the extract below from the report.

"Another recent report from British Marine reveals that 30 per cent of companies have identified critical skills gaps that are holding their businesses back, with 74 per cent of firms citing a lack of technical training as a barrier".

It could be said that future maritime businesses are looking for people with the relevant up to date skills, although I would contest that all skills learned old or new give one a better overall broader aspect and outlook in how ships are built and how to orientate around a vessel, with such hard skills such as glass-fibre replacement of component parts on such restorations as SS GREAT BRITAIN being very relevant to any present day process in the industry and in some respect ahead of the present industry in techniques used.

I myself have lost count of the amount of times I have had to almost take workers by the hand to show them what part of a vessel they should be working on, just another of the challenges when you bring in established trades from outside of the maritime world, they can eventually fit in and do a good job but they need further training in the ways of shipbuilding as there own skills while transferable are not directly transferable.

The restoration and conservation of vessels to add or sustain the maritime heritage in any country is reliant on retaining the skills required to keep the restoration/conservation going and as such each has to rely on the other, one is no good or use without the other, no skills will soon translate into no ships.

3.1 COTTAGE INDUSTRY TO REAL INDUSTRY

The course on the restoration of historic ships/boats that I have been responsible for setting up and authoring some of the ten in total modules along with another two authors with much maritime experience as well as restoration experience, follows the logical progress of what would be required if someone individual or group was to look into the restoration or to a lesser extent the conservation of an existing vessel.

The course follows many of the challenges which one will be faced with and not least of them funding, and as the course does not only contain information on the restoration of historic ships it covers a somewhat broader spectrum of the field of restoration, we also look at what would be required in the restoration of what some may now call classic vessels along with the possible funding routes and class requirements for both if applicable.

As mentioned in the introduction this course is not meant to be claimed as a be all and end all by any means as it has not been approached with any official type backing, but rather as a collection of learned experiences and knowledge in restoration and shipbuilding along with boatbuilding of the gathered authors and the course attempts to pass on knowledge through the written word with leads on where to go for further advice and practical support while also offering to place any students completing the course with restoration programs to suit there own volunteer skills and requirements.

While setting out the course it also occurred to myself that perhaps more could be done with some funding and directed will towards making this route a more permanent part of the maritime industry in the U.K. with dedicated centres to assist and carry out future ship restoration but carried out as a normal shipbuilding business would carry on such a trade.

With two or three centres around the country based in small shipyard that are now closed or close to closing due to lack of work and a structured environment to carry out such work.

A good example would have been the Albion Dockyard which has now unfortunately all but closed down, could this yard have been kept going with future restoration work if such a conduit was in place in the U.K. to fund and direct such projects to applicable resource centres who could carry on the work while training people for the future of what could be a whole new industry based on a very old industry.

Perhaps it could be possible to take the whole concept of restoration/conservation of vessels and view this as industry with a structure set up to cater for this market. Shipyard centres for the restoration/conservation set up around the country working just as a company or shipyard would work, all within a set up that supports the industry with allocated grants for the funding of such and these individual centres, you find a vessel that needs in your opinion restoration and you go to your nearest "Restoration Yard" and pitch your case along with business plan to the centre and they evaluate and make a decision on with the view that for them to continue in the present form they have to have work to do, keeping men in work and creating apprentices with the work there for then to move into once there four year apprenticeship is completed.

The centres could all be run on a semi-commercial basis and funding from the same source can be allocated on a case by case basis although the allocation of funds would have to be made very much quicker, as the "Yard" needs decisions made almost there and then to go ahead with work and such, this would also have the added bonus of opening up this industry to far more people giving much more public awareness to a segment of the maritime world which is not very well known outside of a small section of the marine world.

While is has to be said that Historic Ships U.K has been doing a fantastic job up to now, with there recent initiatives to help grow such centres as described above, and I believe there are three or perhaps four such centres around the country receiving much needed funds to train some boat-building apprentices.

I see nowhere in any of these centres that caters for what will in the future be the majority of historical ships made from iron and steel, the centres are all small boat building based and run and operated by small boat builders.

There is no doubting the fact that the skills base has diminished to a level that could not have been imagined even forty to fifty years ago when I tell you that the shipbuilding industry and all it associated service industries in the U.K. alone accounted for the employment in excess of some two million people
(It is now down to around approximately less than one hundred thousand in the U.K)

The fifty years is quite significant as this is the age of vessels or at least the preferred age of vessels to be considered as being significant enough to be on the historical register of ships in the United Kingdom, and although there are some that are not quite as old as this on the register it is significant to this author that there is not one single ship from the age of the merchant fleet, no fine cargo vessels, no refer ships, no merchant ships at all from the past seventy five years in a time when the design and build of ships took massive technological steps moving from the days of riveted construction powered by steam to all welded ships powered by diesel or even electric diesel and so on.

Along with no centres at all to train or pass on the type of skills required to restore or conserve such vessels, and this would seen to represent a glaring omission from the market of vessels that are looked at for restoration purposes, with perhaps a slight weighted bias towards the old traditional wooden boat, which still has many participants well trained and qualified around the country.

Instead of the restoration of ships and boats being looked at as some type of "cottage industry" perhaps some vision and will needs to go into turning this into a real industry backed by government to create jobs and sustain jobs by the continued maintenance of such vessels which could be based all around the country.

In time everyone gravitates to the water no matter where the live in the country and even with my own passion for the sea and ships if there is nothing happening on the water then it can just become a bit boring, but if there are ships there being worked on in different stages and different types of vessel using skills with people doing and working in these trades that perhaps haven't been seen outside of a shipyard for a good number of years then this can in my humble opinion create an atmosphere that is interesting and more and more people could be attracted to these places.

You only have to look at what happens to ports who hold such as the "Tall ships Events" to see the interest that this can create, but once the ships all go, what happens then?

Surely subsidising such an industry must be a better way than governments just paying out benefits to some of the less wealthy areas of a country, there is no lack of ships to be restored and some could even be restored to help alleviate some of the crises we now face with the huge movements of populations from some of the worlds trouble spots.

An industry that brings together all involved and interested in the preservation of our maritime heritage would also allow all the many enthusiastic volunteers to help and some of the more skilled ones to pass on the knowledge and thereby giving them a real interest in life at a time when the world begins to forget about them and what they have to offer by way of knowledge, that you will not find in a book or on the internet.

Volunteer work helps build critical competencies, knowledge, and skills and can provide valuable training and experience that translates directly to paid employment.

The benefits that can arise from the preservation of ships and boats are many from static exhibits to the centre piece of a museum to being the centre piece to a regeneration project to bring new life not just into the ship but to the local community involved with the restoration.

If the vessel can also move and go out to sea once more then the benefits can be extrapolated even more giving training in the use of such ships to extending experiences otherwise unattainable to people who may be less fortunate than the majority, as said there are no lack of opportunities, I believe that the U.K. is blessed with more than enough experienced and motivated people to do such, providing the will from the powers that be along with the funds are there, this would provide a mechanism to carry forward the preservation of our maritime heritage.

Figure 2: Stern Flag on HMS WARRIOR

A flag that was once seen in every port in the world as seen on the aft end of HMS WARRIOR as she lay in Hartlepool (Photograph supplied by Mr D. Lilley and shown here by permission)

4. CONCLUSIONS

"Preserving the old heritage skills should be just as much a part of our government's responsibilities as the preservation and conservation of old ships"

When I speak of government I speak not just of direct intervention and help but also include in this all the interested parties such as academia and museums, along with such esteemed institutions such as presented before me today at RINA, all sources that could I feel, be very much more involved in the preservation of our overall maritime heritage perhaps it is time to call a ship a ship and dispense with the argument over what should be done and argue about what can be done in a concerted effort for standardisation in what could become a growth industry.

I call this fascinating subject of the restoration or preservation of ships an industry as I firmly believe it to be so and there is also I believe much scope in the furthering of this industry if we can all get together to drive such a positive growth forward.

Rather than have that entire accumulated restoration/reconstruction experience just walk out the door so to speak would it not be better to try and retain such as the Albion Dockyard as just such a centre.

If by helping to produce a course on this subject which is accessible to many at a reasonable cost we can attract more interest from willing people and with luck some of the younger people with an interest and this helps them to get involved or even sparks off a dream then I consider this a worthwhile way of passing on relevant knowledge as "knowledge not passed on is just lost".

I would be happy to discuss further any interested parties in working on ways of driving forward this great part of our proud maritime heritage.

5. DISCLAIMER

Opinions expressed in this paper are solely those of the author.

6. ACKNOWLEDGEMENTS

The author would wish to acknowledge the help of Mr Ian Biles of the Maritime Training Academy for his insight into the subject, and to also acknowledge the help of Mr Derek Hinds of the Trincomalee Trust who along with Mr David Lilley gave kind permission for the photograph of HMS WARRIOR to be used in this paper.

7. AUTHORS BIOGRAPHY

Ron Neish holds the current position of freelance author and shipbuilding consultant/marine surveyor at Maritime Services International. He is responsible as Director and author for courses such as Building your own Boat and Restoration of Historic ships/boats along with numerous other publications for maritime Training Academy. His previous experience includes shipbuilding consultancy on the restoration project for the SCOT II along with more than forty five years in industry in the design fields of automotive, aircraft and shipbuilding, being a time served Loftsman to trade having worked on some forty plus new build vessels from 58 ft fishing craft to the new Aircraft Carriers for the Royal Navy.

His freelance writing work has also been published in maritime magazines on both sides of the Atlantic, while also running and operating his own website on the history of the ships built at Leith.

PRESERVATION OF HMS VICTORY: PEST MANAGEMENT TECHNIQUES

D McCormack, National Museum of the Royal Navy, UK

SUMMARY

This paper will describe the building of a pest management strategy for HMS Victory, after issues encountered with wood-boring insects, including Death Watch beetle. This is crucial to the conservation of the ship, which is an extremely complex environment, and recognised as part of the national heritage. The paper will discuss the long-term challenges of conservation, including factors such as control of environmental conditions within a historic ship in dry dock, providing access to the public as a visitor attraction, and balancing preservation against the ship's status as a commissioned Royal Navy vessel. Various treatments have been carried out, including freezing, heat treatment (Thermo Lignum process) and the use of boron-based applications. Studies are now being carried out to measure the effectiveness of the various treatments in preventing re-infestation of historic fabric, as part of an integrated approach to pest management.

1. INTRODUCTION

In the last year, a new Historic Ships team has joined the National Museum of the Royal Navy (NMRN), concerned with the care and conservation of HMS Victory. As the ship is undergoing a large-scale programme of conservation works to ensure her survival for the next half-century, NMRN is introducing a new approach to her care and preservation. As part of this, the team have been assessing the current condition and environment of the ship, and identifying issues that need to be addressed.

It became clear that there are several pests to historic fabric that are present within HMS Victory, and that these pose a real threat to her long-term preservation. These are well-known pests in museum environments, including wood-boring beetles. Means of eradicating these pests are complicated by the situation and environment within which the ship sits, and several historic attempts at pest treatments have been frustrated by re-infestation. In order to tackle these issues, the team are now adopting a museum-based approach, known as Integrated Pest Management (IPM). This strategy is now commonly used in museums and historic buildings as part of conservation best practice, but is so far relatively limited among historic vessels.

2. HOW TO IDENTIFY THE ISSUES

The first step taken was to conduct a full examination of the ship, documenting historic and active pest activity. The difference between historic and active infestation is not always obvious, so a network of blunder traps around the ship was utilised to harvest a representative sample of all crawling insects present on the ship. The spring season, April to June 2016, was the key period of observation, as this is the time of year when most pests are active, or likely to emerge from the historic fabric as part of their breeding cycle. Hanging traps were also used to catch flying insects, such as moth. Objects within the ship and the displays were checked for pest damage/infestation, and a list compiled of the most at-risk objects and materials among these.

It is also necessary to understand the environmental conditions within the ship in order to illuminate the reasons for pest activity. The system of capturing data on the temperature and humidity throughout HMS Victory has been improved, and these data give us an accurate picture of the on-board conditions and how they may facilitate pest activity. A network of Tinytag data loggers has been activated to record observed values at ten-minute intervals on a 24/7 basis.

Finally, the daily use of the ship as both a visitor attraction and the flagship of the Royal Navy, introduces certain risks to her preservation. These need to be expounded so that the risks can be mitigated.

3. ISSUES IDENTIFIED

3.1 SIGNS OF PEST ACTIVITY

The survey of the ship's interior produced evidence of several pests that were active, or recently active, on board.

3.1 (a) Death Watch beetle

Death Watch beetle [Figure 1] has been known to be present on the ship since at least 1932, soon after her dry-docking. A significant amount of damage can be seen on oak timbers throughout the ship, although in the past, much of the damaged timber has been removed and replaced. Monitoring blunder traps during the 2016 spring season revealed that the adult beetle were present and active on all decks of the ship, constituting a major threat to her preservation. The larva of the beetle feeds on the timber and finally emerges as an adult beetle only to breed, but as the larval life cycle of the beetle can be up to 15 years in length [1], a significant amount of damage is done before the presence of the beetle becomes known.

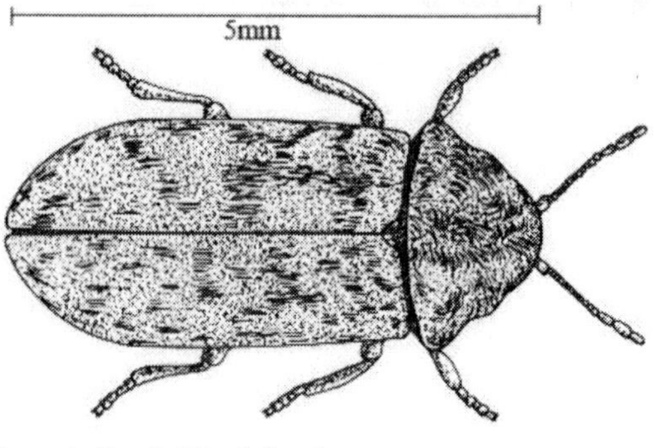

Figure 1: Death Watch beetle.

3.1 (b) Longhorn beetle

A localised infestation of longhorn beetle was identified in the magazine barrel displays in the hold of the ship. This was related to sapwood that had been used to make the hooping on the barrels to create an authentic display. This points to both a lack of quarantine procedures as well as a failure in the housekeeping, as the infestation was not identified until significant damage had been done by the beetle.

3.1 (c) Moth

Webbing clothes moth was found in several different areas within the ship, concentrated in the areas with display dressing such as the hammock spaces and cabins with carpets and uniforms. While it does not directly threaten the fabric of the ship, moth infestation is an indicator of prolonged high humidity and a lack of good housekeeping. It also threatens the display material inside the ship and can be costly to the museum as a result, as material must be treated or replaced.

3.1 (d) Birds and rodents

The dockyard location of the ship presents known risks of rodent activity and nesting birds. Despite the fact that the ship remains open to the exterior environment, there was no rodent damage found during the inspection. One dead rat was found in the hold, but indications were that this was an isolated individual, and no nesting rodents have yet been found. Birds do nest in the masts, which creates mess on the upper deck and a risk of the presence of mites, beetles and other scavenging species entering the ship.

3.1 (e) High risk objects

Other potential food sources for pests within the displays on the ship include a skin drum, leather, feathers, woollen blankets, hair-stuffed mattresses, ropes and sweets in the shop stock. These are potential attractants for common museum pests such as carpet beetles, biscuit beetles, mites and rodents.

3.2 ENVIRONMENTAL CONDITIONS

HMS Victory sits in dry dock in Portsmouth Historic Dockyard. It is open to the elements and adjacent to the harbour. She is open to the public in all weathers, welcoming visitors all year round. She is therefore sensitive to rain, wind, ice, snow, spores and moisture carried in by visitors.

The temperature and humidity data collected on the ship over the last year show a trend that is, unsurprisingly, not ideal. The data demonstrate that, unlike in a closed system, relative humidity (RH) and temperature are not in a directly proportional relationship [Figure 2]. Humidity remains much higher than is desired throughout the ship and all year round, indicating that the timber structure itself is holding onto more moisture than is recommended. With an average RH in the region of 72%, the moisture levels in the ship and its environment are key to understanding many related challenges that she faces, including pests and rot.

Although the ship is not tightly sealed, internal airflow is poor, resulting in a stagnation of the environment. This is more pronounced in the lower decks and hold, where there is a corresponding rise in average humidity.

Mould and fungus spores are omnipresent and will thrive in conditions above 60% RH. The timber and other organic materials on board the ship make for nutritious substrates upon which many fungal species will thrive. These include species that will degrade lignin and cellulose, and are therefore a threat to the structural survival of the ship. Where the RH remains constantly high, the mould and fungus present can become established, grow and reproduce unchecked. Reducing the RH within these areas will slow or halt the growth of these types of organisms, but will not in itself prevent them from becoming active again in future. In conditions of poor ventilation, it becomes even less likely that airborne spores will be naturally cleared from interior spaces.

There is also evidence that fungal decay allows attack by Death Watch beetle on oak heartwood that would otherwise not be viable as a food source. Reports from the HMS Victory quarterly reviews [2] in the 1970s show that at the same time the beetle population counts were showing a sharp rise, there was also a notable spread of fungal attack around the ship. This suggests a symbiotic relationship, wherein the beetles act as carriers for the fungal species which also promote the decay that allow them to feed, grow and reproduce.

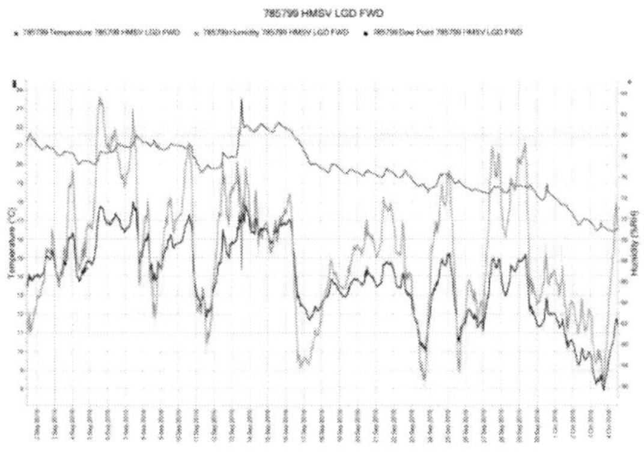

Figure 2: Showing typical data captured on the ship.

3.3 THE HUMAN THREAT

3.3 (a) Museum visitors

On average, HMS Victory welcomes 400,000 visitors on board every year, and is open as an attraction 364 days in the year. While the visitors are vitally important in sustaining the care of the ship, through public interest and admission fees, they also bring aboard certain risks.

Visitors affect the environment within the ship by bringing moisture inside, not just on wet weather days, but simply through the action of breathing. They are also carriers, albeit unknowingly, of fungal spores and potentially even insect life. Facilitating safe access to the ship also requires leaving the ship physically open to the outside environment and the dock bottom, providing more potential access points for pests.

There are also events which take place on the ship that include large dinners, which underline the need for rigorous housekeeping. The interior of the ship is cramped and dimly-lit, and it can be a challenge to ensure that all events pass off without any detrimental effects to the ship.

3.3 (b) Naval presence

As the flagship of the Royal Navy there is a constant naval presence on board the ship, and those officers who work on board must of course have facilities to eat and drink. While cooking is not permitted on board, there is still an inherent risk that the presence of food, if not carefully controlled, will attract pests including rodents.

Certain naval ceremonies take place on board throughout the year, and many of these utilise fresh flowers or wreaths, in commemoration or celebration of different historic events, such as the Battle of Trafalgar. This also constitutes an attractant and a possible mode of entry onto the ship for insect pests, and it is important to introduce procedures for managing this material to minimise risk.

4. HISTORIC APPROACH

The traditional approach to dealing with pests on HMS Victory has been a reactive one, i.e. when infestations were discovered, treatments were carried out.

When the presence of Death Watch beetle on the ship was initially discovered in 1932 it was recognised as a serious threat to the oak structure. In 1937 a regular recording system was put in place and the numbers of beetles caught was recorded on a yearly basis from this point forward, with a few interruptions during the 1940s [3]. Beetles were physically collected, several times a day, from the end of March until the end of June each year. Some of the early treatments that were attempted included application of liquid insecticides through bore-holes in the ship's side, supplemented with DDT dusts on the decks [4]. In 1954 the decision was taken to seal and fumigate the entire ship with a methyl bromide gas, a process that was repeated in 1955 and 1956. The result of this was a drastic reduction in the number of beetles captured during the annual harvests, from 5,200 in 1953, down to 72 in 1957. Nevertheless, the beetle survived and continued to spread once more throughout the ship. By 1978 the count recorded 2710 beetles, and it was at this point that an annual lindane smoke treatment was introduced, alongside an annual spray application of permethrin; this continued until the late 1980s but during these years the beetle numbers recorded continued to rise. The system of beetle survey and capture was then suspended for some years because of major restoration works, and has never been resumed in the same manner.

In more recent years, baited blunder traps were employed to record Death Watch beetle numbers, but no system of recording all species encountered on board has been devised before now. When an area was seen to be heavily infested with Death Watch, a boron-based insecticide was applied to that area. The boron solution is very effective in eradicating insects and larvae, but is limited by the extent it may penetrate into the substrate, so the success of the treatment can only be judged by consequent observed numbers of beetles in the traps.

These attempts at pest management are all hindered by their reactive nature, and cannot be truly successful without a deeper understanding of the root causes of the issues. For this reason, it was decided that a new approach was necessary that could reveal the baseline conditions and populations present, but also show how they were related to each other. This should illuminate the factors we can try to manipulate to improve the conditions. It is hoped that this will allow a more preventive strategy to be adopted, and a comprehensive programme of regular monitoring to be introduced.

5. THE INTEGRATED APPROACH

5.1 WHAT IS IPM?

The impetus behind the integrated approach is to be fully informed and be able to both suppress the pest populations on the ship and to prevent any re-infestation. The goals of utilising the IPM are as follows:

- To suppress the pest populations in the historic environment and fabric of the ship, both insect and fungal.
- To reduce the cost and toxicity of interventive measures that are necessary to solve current and future problems.
- To instigate preventive measures that protect against any further infestation, by exercising greater environmental control.
- To ensure that all persons using the ship take responsibility for the IPM in an appropriate capacity.
- To use all expertise necessary to devise a comprehensive monitoring system that will track the progress and success of the programme.

5.2 HOW DO WE START?

After the initial survey work to identify the issues on board, an IPM strategy was drafted that was specific to the ship and her challenges. The document outlines persons with responsibility within the strategy, and is then circulated to all relevant teams in the museum and Royal Navy. For example, the museum's commercial team are responsible for ensuring that all consumable stock in the shop is sealed and stored off the floor, and that all rubbish is cleared from the ship each evening at close of business. The document also contains annotated deck plans of the ship which show positions of traps and known areas of pest activity. Overall responsibility for the strategy and its review is held by the Conservator of Historic Ships. One of the key factors in the success or failure of the IPM is documentation; unless the pest activity, treatments and monitoring is recorded and brought together in a usable manner, the strategy will not be effective. The IPM document discusses measures to be taken as described below.

5.2 (a) Housekeeping

As part of the process of introducing this IPM to the ship, the museum has brought the cleaning contract for the daily care in-house. A new team of Ship Keepers is cleaning the visitor route daily, but also undertaking a deep clean of the entire ship to improve the baseline conditions. The team is being trained in basic conservation and collections care methods, including how to recognise pest activity, damage and frass.

5.2 (b) Environmental control

A robust environmental monitoring system is now in place and we continue to harvest data on the relative humidity and temperature inside the ship.

There is a dehumidification unit in the aft end of the Orlop deck, and the data demonstrate that it produces around a 10% reduction in the RH in this area. This is sufficient to hold the average RH in this area below the 65% at which mould and fungus becomes most prolific. It is also currently one of the only areas within the ship that shows no active Death Watch population. These are encouraging signs that artificial environmental control will help us to suppress biological attack. As part of the new procedures, regular moisture content readings from timbers throughout the ship will be recorded, to help track progress against leaks, humidity and rot.

The data captured will now inform our choices in placement of other units to try to drive down the humidity in other parts of the ship.

5.3 (c) Removal of food sources

Display material will be examined and where possible adapted to remove all potential food sources. As mentioned above, sapwood hooping on barrels allowed infestation by longhorn beetle, and this material has now been removed from the ship. Other materials that can be removed without compromising the presentation of the displays include: the hair stuffing of old mattresses, which can be replaced with synthetic stuffing; straw in the animal pens, which can be replaced with plastics; and, sheepskin on gun sponges, which may also be replaced with an inert synthetic substitute.

5.3 (d) Remedial treatments

In order to deal with the known current pests on the ship we have several methods at our immediate disposal.

Freezing infested objects has been employed on textiles and barrels found to contain moth. This has been carried out on site by Mary Rose Archaeological Services and is very effective in killing most pests. Objects must then be thoroughly cleaned before they can be re-used.

The jeer capstan was treated by Thermo Lignum in 2016 as this was the only treatment available that was capable of destroying all stages of the Death Watch beetle in an object of this size. The capstan is more than a metre in diameter and evidence showed that the beetle were active at the very heart of its structure. This resulted in many visible exit holes and significant loss of material on the outer surface of the structure, as shown in the photograph below [Figure 3].

Figure 3: Showing typical beetle damage to the capstan.

The capstan had to be treated *in situ*, and so was enclosed in a treatment tent [Figure 4]. The temperature within the tent was raised to 55°C while the RH was controlled at 58% [5]. These conditions are hot enough to destroy enzymes that are crucial to the insects' survival, but the stable humidity protects the wooden capstan against any warping or splitting from moisture content changes. This treatment had the added advantage of being entirely non-toxic and so could be carried out without closing the ship to the public.

Figure 4: Showing the capstan under treatment.

In addition to these measures, the application of a boron-based biocide is continuing wherever new growth of fungal bodies or new Death Watch activity is encountered.

5.3 (e) Preventive treatments

In order to prevent infestation of those materials that remain potential food sources for pests on the ship, such as textile dressing materials in displays, a spray-on application of Constrain insecticide (registered under the Pesticides Regulations Act of 1986) has been selected for use on the ship. This product is pH-neutral and safe to use on most materials, and when dry is safe to be handled without any health risks. Objects are treated in the museum work areas and then replaced on the ship once dry. Constrain can also be applied by brush and is effective against most museum pests, so can be topically applied to furniture to protect against wood-boring beetles.

It will also be important to extend the principles of this IPM strategy to the timber stores that hold the 'new' materials used in repairs on the ship. These stores are now monitored with a network of traps to identify any pests present so that issues may be dealt with before this material reaches the ship itself. Options for pre-treatment of timber with biocides are now being discussed as part of the preventive strategy.

5.3 (f) Quarantine

Quarantine space on site will be provided that may be used to transition objects on and off the ship safely, to ensure that no infestation is carried on board. Stores are currently being assessed so they may be adapted to provide better conditions for this; ideally quarantine rooms would be uncarpeted, with a sustainable mid-range RH, adequate light to inspect objects within, and secure doors that provide a complete seal to the room. Racking and means of storing/packing objects off the floor would also be required. As with many museums, working within historic buildings provides its own challenges in terms of ideal working spaces.

6. FURTHER RESEARCH

The IPM is a strategy rather than a fixed document and will of course be adapted and updated according to the progress made on the ship. There is still much work that can be done to shed light on different aspects of the issues discussed above, particularly the relationship between fungal decay and Death Watch beetle attack, and the selection of a method for quantifying damage in the ship's timbers. Some of the areas of interest are discussed below.

6.1 ACADEMIC PARTNERSHIP

Feasibility studies have been initiated with other bodies, including involvement with student research projects, to study factors that are key to future conservation efforts. These include a study into factors affecting oak degradation in ships and potential methods of nondestructive quantification of beetle damage in timber. Fungal species identification is being carried out, and the links between insect carriers and cellulose-degrading species are being investigated alongside a conservator from Birkbeck College, with analytical assistance from Historic England. It is hoped that findings from these research strands will allow us to find the most effective way to eradicate the decay and associated pest attack on HMS Victory, while also allowing us to instigate an efficient and informed set of monitoring procedures to prevent future biological deterioration of the ship.

6.2 ENVIRONMENTAL MODELLING

There is still work that can be done within the ship itself to allow improvement of the internal environment. We have identified that there is a lack of good ventilation in the ship which works against the drying of the timbers and encourages mould growth. It is anticipated that airflow modelling would allow us to make best use of artificial environmental control including dehumidifiers, heating units and fans. These measures are non-invasive and therefore to be encouraged, but if used incorrectly they are inefficient and not cost-effective to run. For this reason, we want to combine modelling with environmental data capture to ensure that we can have the most effective impact with the minimum energy input necessary.

7. CONCLUSIONS

There is still much to be done in eradicating pests from HMS Victory, but the IPM approach described here allows more comprehensive and informed methods of pest control to be adopted. By striving to understand the very complex environment of the ship, we can begin to understand it as an ecosystem, and therefore to tackle the problems of humidity, mould and fungal growth, and insect activity at the source.

8. ACKNOWLEDGEMENTS

The author would like to acknowledge the kind support and advice of IPM specialist David Pinniger, Conservator Sophie Downes of Birkbeck College, Historic England, Thermo Lignum, Dr Fiona Brock and Richard Southwell at Cranfield Forensic Institute, colleagues at NMRN, the Mary Rose, and the crew of HMS Victory.

9. REFERENCES

1. Pinniger, D., *Integrated pest management in cultural heritage*, 2015.

2. General Manager, Portsmouth Dockyard, HMS Victory Advisory Technical Committee Report No.44 (*unpublished*), 1976.

3. McGowan, A., *HMS Victory: Her Construction, Career and Restoration*, 1999.

4. Bugler, A., HMS Victory: Building, restoration and repair, 1966.

5. Mizuno, J., Treatment Report: HMS Victory, Portsmouth 2016 (*unpublished*), 2016.

10. AUTHORS BIOGRAPHY

Diana McCormack holds the current position of Conservator for Historic Ships at the National Museum of the Royal Navy. She is responsible for the ongoing conservation of the museum's collection of historic vessels, including HMS Victory, HMS M.33 and the submarines Holland I and HMS Alliance. Her previous experience includes conservation of large objects and engineering collections at the Science Museum and National Museums Scotland, as well as archaeological conservation for Historic England (formerly English Heritage).

THE BANTRY BOAT – RESEARCHING & CONSERVING A UNIQUE HISTORIC VESSEL

J Kearon. AssocRINA UK. Retired: National Museums Liverpool (Merseyside Maritime Museum) UK; National Museum of Ireland.

SUMMARY

This paper will outline a brief history of the vessel known as *The Bantry Boat* and of the circumstances that led to its capture in Bantry Bay, County Cork, Ireland in 1796. It will also provide a concise overview of the Atlantic Challenge organisation and the uplifting work it has achieved internationally over the past 30 years. Not least it will outline the vessel's built structure, its condition pre–conservation, its provenance and the conservation approaches taken in preserving this unique vessel which is some 90% original in form, but has undergone considerable neglect and damage over time. It will also outline the methods taken in physically accessing and working on the vessel and some of the innovative approaches taken to treat and protect its structure. But most significant will be the related research in trying to assess the vessel and its history after capture, and the eventual success in establishing some surprising but important findings that emphasise the hidden information that can be gleaned from unique and significant historic vessels.

1. INTRODUCTION

1.1 THE FRENCH INVASION 1796

Known as *The Bantry Boat*, given her place of capture, the vessel is an Admirals barge from the French frigate *Resolue,* of 1778, flagship of Rear-Admiral Neilly, and one of 45 ships that sailed from Brest in December 1796, with 12,000 soldiers, to invade Ireland. The United Irishmen rebel leader Theobald Wolf Tone was with the fleet with the intention of joining fellow rebels and with the invading French, to provoke an uprising to take Ireland by the back door, so to speak. The expedition was a failure, predominantly related to bad weather and poor command, with ships being dispersed by gales. Had the expedition been successful, it could have changed the course of European history.

Following a collision off Bantry Bay between *Resolue* and ship-of-the line *Redoubtable,* the barge was launched under the command of Lieutenant Proteau, to seek assistance. Engulfed in atrocious weather, she was beached and her crew was arrested by local militia, under the command of Richard White, (later the Earl of Clodagh). He brought the vessel to his home, Bantry House, where she stayed until 1944, when presented to the National Museum of Ireland (NMI) by a descendant of the Earl. The Barge was placed in storage by NMI, and remained so until 1972, when it was loaned to the voluntarily managed National Maritime Museum of Ireland, Dun Laoghaire. The Barge remained there, indoors on display until 2001, when for structural reasons the museum was closed.

1.2 THE ATLANTIC CHALLENGE PROJECT

Members of RINA will largely be aware of the *Atlantic Challenge Project* and of the associated sailing / pulling gigs, built internationally [1]. The Project was founded in the mid 1980's by Lance Lee, of Rockport Apprentice Shop, Rockport, Maine, USA. Lee was influenced by the Outward Bound School's philosophy of Kurt Kahn and Lawrence Holt. He consulted with colleagues, including Bernard Cadoret, editor of ***Le Chasse Maree,*** in searching for the ideal prototype vessel for a match-boat contest that could be built, then rowed and sailed by the young people who built them. They searched widely both in Europe and USA. After viewing the *Bantry Boat*, she was chosen as the model on which the Project's boats would be built. To date some 70 exact replicas have been built, which compete in sailing and rowing competition, with the young people that built them, also participating in their use

Figure 1. *A 'Bantries' gathering at Bantry, July, 2012*

1.3 CONSERVATION

Following the closure of the National Maritime Museum in 2001 the need to relocate the boat prompted NMI to consider its future. Having been involved, through my position with National Museums Liverpool, in analysing the historic yacht *Asgard* for NMI, I was asked to also examine the *Bantry Boat*. This resulted in NML taking on the contract to treat and conserve the vessel in a museum conservation context with the emphasis placed on protecting, consolidating and conserving as much of the vessel's original structure as possible. This approach would also extend to the vessel's coatings, which were largely assumed to be original.

In 2002 the vessel was transported from Dublin to the Department of Maritime, Engineering & Transport Conservation at Merseyside Maritime Museum (NML) for full assessment. During 2005-6 the barge underwent a detailed programme of work combining research, conservation, restoration and replication, where appropriate, eventually being returned to Dublin and placed on display at NMI, Collins Barracks. The NMI had virtually no information on the vessel related to provenance or usage. In consequence the surveying of the vessel and related research within the Department was intensive and thorough, given the vessel's status, with a full and in-depth record of the entire project, including coatings and materials analysis, an intensive structural survey, photographic recording and detailed records of everything that was done, held in NMI Archives. I was ably assisted by my colleague Jonathan Carr, Museum Ship-keeper and Historic Vessel Conservator, in particular related to the complex structural replication aspects, which he undertook throughout the project.

2. STRUCTURE & CONDITION

2.1 HISTORIC PHOTOGRAPH

A black & white photograph of the vessel on land beside a tidal inlet at Bantry (date unknown) shows the barge seemly in good condition with no apparent damage. The image was referred to in a letter dated 29th April 1947 from the Art Division of NMI to Commander Stuart Jones, RN, Somerset, England, stating the photograph was taken 'many years ago as she lay at Bantry'. It is the only clear view of the vessel at Bantry. No images appear to have been taken from the time the barge was taken into NMI's Collection in 1944, when damage and missing components were referred to in the same letter.

Figure 2. The *Bantry Boat* by an inlet at Bantry, date unknown

Figure 3. The barge after arrival at the Shipkeeping Conservation workshop, Merseyside Maritime Museum

2.2 STRUCTURE

The barge is an open wooden carvel vessel of Scots pine on sawn compass elm frames. The centreline structures of stem, apron, sternpost and stern knee, also of compass sawn elm, with a transom of elm planks and a keel of rock elm; the vessel has no hog or keelson, the garboard planks being butted to the side of keel, but with no rabbet. The stem is half-jointed to the fore end of keel and backed by a two-component apron. The sternpost is badly damaged with its bottom portion missing. The keel likewise is damaged over its length, with five breaks and some three metres of its after portion broken off, and some 800mm of the aftermost piece missing. The stern knee is badly degraded.

There are 11 strakes each side, all of 18mm in thickness. Both the sheer strake and gunwale are formed and fixed in clinker construction fashion, with the remaining nine strakes formed and fixed in standard carvel fashion. The carvel strakes are fixed with square-shank rose-head bronze nails, the sheer strake and gunwale are fastened with round iron bar fastenings that are peened over copper roves within and without. The majority of the strakes are formed from single planks. The Scots pine was unexpected and more likely to be found in a Scandinavian built vessel, and with random knots and shakes throughout. However the majority of the planking was in relatively good condition structurally. The yellow pine forming the after quarter decorative side panelling also showed knots and shakes.

2.3 OAR PORTS

The gunwales each have five alternating oar-ports in the form of strengthened rectangular openings built into the gunwale, with No 1 thwart serving the first oar-port on the starboard side. Each port is lined with leather cladding held in place with copper tacks. The leather and fixings appeared to be original, given no evidence of previous repairs. Most of the leather claddings have sustained some damage, from the mild to almost total loss, with severe loss of leather from most of the port side oar ports. On the starboard side the opposite was found; only oar port No 7 had lost its cladding, with the remaining four starboard ports being largely undamaged. This strongly indicates uneven oar usage of a form difficult to analyse. With average usage in such a vessel wear-and-tear would be relatively evenly distributed. A theory put forward is that in the century and more after capture, she may have been used

for seine-net fishing. Such can involve circular rowing that over time can cause uneven wear in rowlocks.

2.4 FRAMES

There are 33 frames, the majority formed of a single sawn floor centred on the keel and overlapped by futtocks that extend from above the keel to butt to the shelves that run parallel with the top of the sheer strake. The frames are fixed in place with a single iron boat-nail through the centre of the floors and into the keel. The frames and floors of elm were of good quality and in the main had weathered relatively well. However, the central portion of the majority of the floors - specifically in the area above the keel - had degraded, with many rotten. This had much to do with the corrosion into scale of the iron fastenings.

2.5 THWARTS

There are 10 thwarts, the majority being of birch, a wood not usually found in wooden boats and prone to insect attack and rot in the open. Each of the thwarts shows evidence of previous wood-worm infestation, in varying degree on their undersides. However, thwart No 2 was of elm, but also infested. Thwarts No1 & No10 were missing. Thwarts No2 & No6 are both rigged and strengthened to support masts, with hinged iron collars fitted to their after edges. There are also shaped widening pieces fitted. Thwart No 5 also has evidence of a central mast collar being fitted in the past, now missing. This thwart is different in some respects to the other mast-support thwarts, being of a single plank, but with no widening pieces fitted. These significant aspects of thwarts No's 2, 5 and 6 will be dealt with in Section 9: RESOLVING INCONSISTENCIES.

The thwarts rest on a stringer and are secured in place by a single grown elm knee to each side. Wooden seat pads, similar in section to the thwarts, with rounded inboard ends, are fitted alternately to one side of the top of each thwart, where the oarsman sits. The pads are notched around the thwart knees, enabling oarsmen to sit snug against the vessel's side when rowing.

2.6 AFTER QUARTER ARRANGEMENT

The after quarter of the boat from Thwart No 10 to the transom was originally fitted with side seating and a raised level of decking above the keel to suit; these structures were missing from at least the time the vessel came into NMI care, correspondence of the period prove. The side seating positions are backed by original panelling (still in situ) decorated with elongated rhombus-shaped crests defined with fine wooden beading, four sections to each side, each diminishing in size from where the panelling butts to the transom. A single large crest, to match the side versions, is outlined on the inner face of the transom,

3. ASSESSING DAMAGE

3.1 Cause of Damage

The cause of the structural damage and loss of material to the sternpost, the keel, stem and garboards is unclear, but suggests accident or misuse, with some corrosion related damage to the joint of stem and keel and also much related to floor fixings. The loss of thwarts, after sheets seating is unexplained, but possibly related to neglect and rot. The damage to the floors is directly related to exposure and corroding iron fixings. The old black & white image of the boat shows no such damage. Given the evidence, it becomes obvious that neglect and exposure over time was the cause of the rotten floors and the extreme loss of coatings throughout the bilges and on the thwarts. A clear case of a boat left unprotected in the open.

3.2 COMPLEX PROCESSES

From missing sections of batten within the gunwales to the more complex consolidation of damaged decorative panels, the smaller items were relatively easy to deal with. The more complex processes of restoring structural integrity to the vessel as a whole, while trying also to save as much original material as possible, certainly focused our minds. In this paper it will be these aspects in the main that will be outlined in some detail to provide insight of the reality in trying to consolidate and conserve a unique but much damaged object.

3.3 MOST DAMAGED ASPECTS

The principle structural problem identified in the vessel was the damage to the keel and floors and the loss of transverse connectivity. The keel was central to this and its removal became necessary to deal not only with its damage, but to try and consolidate it in some way to give the supportive strength required. But dealing with the damaged floors became the real issue, given the extent of damage involved and their central role in transverse integrity. The damage to stem/keel joint and the lost lower sternpost also required attention, given its central role in connecting to the keel. Some replacement/replication of missing and damaged material would be necessary. Internal damage and loss included the absence of both No 1 and No10 thwarts, the loss of the after quarter side-seating, also of the related sole-boards and bearers. Such would be relatively easy to resolve, in effect straightforward replication in original form.

Figure 4. A view looking aft, the damaged floors and the space where the after portion of keel should be are obvious.

4. CONSIDERING METHODOLOGY

4.1 SUPPORT

Given the size of the boat, the structural damage, its fragility and the need to access every aspect of the outer and inner vessel and also to remove the keel, it became necessary to design and construct a support cradle that would allow such access. A cradle was formed (within the Department) with five evenly spaced support frame sections, each shaped to the hull specifically and offset longitudinally along the pivoting centres, which would allow the boat to be canted to both port and starboard sides, also to allow the keel to be dropped and removed for repair. The access given meant the vessel could be placed in different angled positions to suit specific requirements. Furthermore, when angled, the upward facing side could have that sides support sections removed quickly to allow full unimpeded access to that entire side and requiring just three people to undertake the move. This proved a boon in time saved throughout the project.

Figure No 5. The barge supported in the canting frame.

4.2 KEEL

The keel was removed in its several broken sections. It was very damaged and degraded, particularly the bottom and in its core. Creating a new keel was considered, however a more innovative approach was taken, bearing in mind the seemingly opposite approaches of conserving original material while maintaining structural integrity. Our approach was to core out the damaged centre of the keel, which was degraded through corroding fastenings, leaving the sides and top intact and inserting a new core. This was achieved by forming a sturdy plywood box, the dimensions inside being the same as the keel, and the same length overall. The broken sections of keel were then inserted into the box, bottom face up, with a new section included to replace the missing after end. Setting up the box to use a router, making several guided passes along the edge of the box, the core was routed out, leaving a U shaped remnant of top and sides, with walls of some 4mm. A full length piece of prime clear Oregon pine was then glued into the shell of the keel, resulting in a strong and solid core within and the original sides and top outside.

4.3 FLOORS

Repairing the floors involved removing them, most were in two sections, given the central rot. But rather than remove the hull fixings from the floors, we very carefully lifted the floors from the fixings. This was to overcome a problem related to outer coatings and countersunk nail stopping, original and undamaged. By doing this the nails remained in place undisturbed, with no damage caused to their countersunk stopping or outer coatings. The two sections of floors (templates had been made of the original full shape) were then laid and clamped to a plywood base and the damaged lower ends cut back to sound wood, with a chock of compass elm matched and fixed in glued-scarph fashion to the floor sections. This resulted in a complete sound floor, the majority being original.

The original nail holes were cored out to a larger dimension, and then filled with a thick epoxy compound and masking-taped over to prevent spillage. The floor was then pushed back exactly in position with the nails entering the epoxy filled original holes, all forced together tightly until cured. The repaired floor was at full strength in position with the absolute minimum removal of original material, by taking advantage of the nature of the rot and angle of good wood, this could not have been achieved by repairing the floors in-situ. When all the damaged floors were so treated, with some variations, the restored keel with its outer original structure was also refitted and the floors re-fixed to it. Full structural integrity was achieved.

Figure 6 A view through the bilges showing repaired floors.

4.4 AFTER SHEETS REPLICATION

The after sheets seating and decking were replicated in exact original fashion, helped by the survival of two of the original floor bearers, one aftermost transverse seat bearer and by a curved support to the mid-section of side seating, all which indicated positions and dimensions. All new wood used in the barge to replicate or repair missing components, or component parts, such as in the floors, matched original species and was left unpainted, being lightly matt-varnished only for sealing purposes; to define such from original build material.

4.5 DECORATIVE PANELLING

The decorative panel beading was damaged to some degree, with several sections of batten missing. Damage was caused principally by corroding tacks securing the beading. These pieces were replicated and the damaged pieces were repaired by the scarping of small pieces of the same wood (yellow pine) into the degraded fastening locations and shaping them to match original form. The process worked well and resulted in the decorative aspects of the panelling being largely restored and being largely original in form.

5. COATINGS

5.1 CONDITION

The appearance of the barge on arrival in our workshop in Liverpool can be summed in one word - Dirty. The entire interior, including thwarts, shelves and stringers, was coated with a layer of ingrained dirt and dust that mostly obliterated the remaining coatings and bare surfaces. The only saving grace was that no oil or grease was present, just ingrained dirt, layer upon layer.

5.2 OUTER COATINGS

The boat was coated with paint on both exterior and interior surfaces, but with coatings much damaged, degraded and lost, particularly in the interior. Four different colours were evident; white, red, blue and dark grey. The outer hull is coated with white paint from keel to below the sheer plank, it being coated with red paint and the gunwale coated with blue paint. The port side in addition was randomly covered with pencilled graffiti, mostly on the fore port quarter, in the context of people writing names and dates, spanning from 1920 to 1944.

5.3 INTERIOR COATINGS

The interior surfaces from below the thwart stringers are coated with a dark grey paint, with interior surfaces of both the gunwale's and sheer planks painted red. Frame tops above the thwart stringers are coated in blue, as are the sheer plank shelves, thwart knees and oar ports. The thwart ends are painted red out to the knee ends, as are the oars-men's seat pads. The central portions of the thwarts were painted blue, though their coatings are now virtually non-existent, further evidence of being out of doors long-term.

5.4 FRENCH NAVY COLOURS

The general belief perceived was that the coatings were original to 1796. But we were dubious about the colours used. We then sought to determine the actual colours used by the French Napoleonic Navy on their ships boats. Through colleague Justin McMorrow, National Maritime Museum, Helene Tromparent, Ship Model Conservator with the Musee National de Marine, Paris was contacted. She in turn contacted Jean Boudriot, the noted French maritime historian, expert on French Napoleonic Navy ships, and author of the definitive publication 'The 74 Gun Ship'. He confirmed that French ships boats of the period were painted white on the exterior, with a black sheer/gunwale and indigo on the interior.

5.5 ANALYSIS of COATINGS

Analysis of the vessel's coatings [2] found that the blue paint on the gunwale and elsewhere contained the pigment ultramarine, but not of the natural pigment (which was extremely expensive) but of the synthetic form. This was not invented until 1828 and so placed the blue paint on the vessel to at the very least, post 1828. The structures of the layers painted red are very similar to the blue areas and these also have been repainted several times. The other coatings on the vessel largely contained combinations of materials that were in use in paints and varnishes generally available in the late 18th Century. However, doubt was expressed regarding the presence of barytes (barium sulphate) in most of the coatings and it was suggested that if this was a material in use in the late 18th century, then some of the paint layers, such as primer and intermediate coatings could be original. In fact, barium sulphate and the pigment ultramarine were also identified in mid19th century coatings on HMS *Victory*.

Figure 7. A cross-section of paint layers showing the artificial ultramarine coatings.

6. SURFACE CLEANING

6.1 CONSIDERING APPROACH

Given the dirt encrusted surfaces of the interior hull and the general surface damage to the outer coatings several options were considered for how best to clean the vessel while minimising damage to remaining coatings. There was no question of removing coatings and freshly repainting the vessel. In consequence a total removal of dirt and grime was decided on. Several options were considered and sampled including laser cleaning and solvent use; both damaged the paint surfaces. However, the most successful was the simplest; using water and cloths – and dedicated elbow power.

The vessel was thoroughly and meticulously cleaned by using water, a combination of several different weaves of cotton cloth, absorbing sheathing and fine sponges and fine brushes (even fine tooth-brushes). By working through several levels of repeated cleanings, the dirt was removed and original coatings and wood surfaces were exposed. It became a revelation.

Figure 8. A first sighting of colour in many years

On completion, the vessel was coated with several applications of Polyvinyl-alcohol, as a protective coating, which was reversible.

7. PROVENANCE

7.1 DESIGN

The basic design of the barge was not uncommon in European waters at the time of the Napoleonic wars; however, it differs considerably from the standard ships boats carried by both British and French Navy ships of the period. In the main these ships boats were bluff bowed and wide beamed, with variations in size and structural strength to suit specific purposes. *The Bantry Boat*, in contrast is a long, slender and lightly built vessel with an extremely fine entry, the opposite, in fact, of standard ship-of -the-line boats. Paul M. Kerrigan, in his paper on the vessel states [3] "It seems probable that this was the barge of Rear-Admiral Neilly --- and was the last boat to be hoisted aboard before departure of the fleet." However, he also surmises that she "may be from one of the 74-gun ships of the rear-guard, or was in use in the harbour of Brest".

7.2 FRENCH PRISONER of WAR MODEL

Some doubts have been raised about the vessel's origins; that she was not the actual boat used in the rescue attempt, even that it may be a British Royal Navy vessel, captured by the French, though such was not unusual. However, this is unlikely, largely based on the type of decoration found backing the after quarter seats. The elongated rhombus form of such is largely believed to be of French origin, while the British Navy, and British Royal and Thames Ceremonial Barges display rectangular decorative panelling [4].

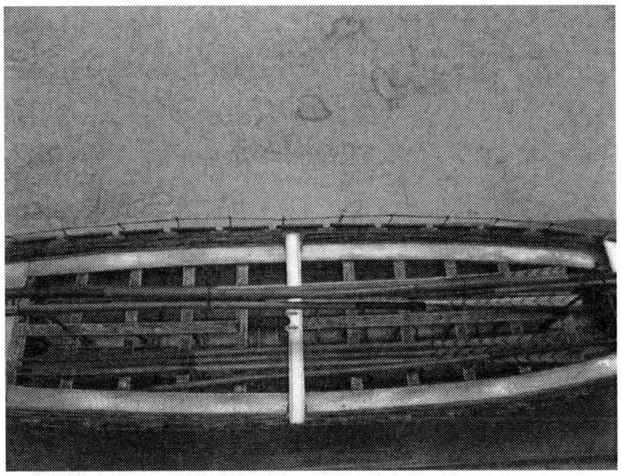

Figure 9. Note the same form of inboard decorative panels on the French model as on *Bantry,* also the same form of single plank mast-bearing thwart as No 5.

The most compelling evidence of the vessels origins however, relate to a French Napoleonic prisoner of war period model of a pulling gig from National Museums Liverpool Ship Model collection. Brought to my attention by Chris Mosely, Ship Model Conservator, this wonderful model barge, while of a different layout, and carrying cannon fore and aft, is very close in hull design to the *Bantry Boat.*

It shows the interior decorative panelling to be both of British and French forms, but with the latter the most prominent. This can be taken to be in the context of gaining a better price for the model [5] by bending somewhat to include the British method in addition to the prisoner craftsman's own country's version. Other comparisons in the model relate to the oar-ports, the same form as on *Bantry.* Furthermore, both of the model's two mast supporting thwarts appear identical in form to Thwart No 5 on *Bantry.*

8. RESEARCH

8.1 BANTRY HOUSE

In an effort to gain some insight or information related to the vessel's usage following capture contact was established with Egerton Shelswell-White, owner of Bantry House and a direct descendant of Richard White, who commanded the militia that captured the vessel in 1796. Shelswell-White said the boat was stored at Bantry House from its capture until 1944 when it was donated to the NMI. He had no information on whether the boat had been used after capture. His earliest memories of the boat date from c.1940 when as a young child he 'played in the boat freely' - it was in the boathouse near one of the estate's gate-lodges. I was then directed to University College Cork (UCC) where the family archives were stored.

8.2 UNIVERSTY COLLEGE CORK

At UCC I found that the Bantry House Estate Office had been badly affected by a fire in c.1900 and very little pre - 20th century records or material had survived. After searching the archive thoroughly – entering words that might produce some information about the boat, such as: - French invasion; 1796, boat; fishing; sailing; shipwright; boat-builder; blacksmith etc. Indeed any word associated with wooden boats or the materials and crafts associated with them. No results showed. There was not a scrap of information on the boat or its use, apart from the known correspondences related to its capture in 1796.

8.3 USAGE POST 1796

There is little information related to how the vessel was treated following her capture, particularly in relation to the 100 years or so up to the beginning of the 20th century. It has been stated that the boat 'went unused for 140 years' [6]. However, its condition reflects much use, damage and neglect, particularly related to some of the oar-ports, keel, and sternpost and after sheets seating. This suggests that the vessel was probably actively used in the Bantry Bay area, in varying degrees over a considerable number of years. Indeed the finding of the artificial ultramarine paint testifies to continual use.

9 RESOLVING INCONSISTENCES

9.1 SYNTHETIC ULTRAMARINE

The principle element that concerned me was the vessel's use during the century and more after capture. No evidence survived apart from her condition and the contrasts between the old image showing her in good structural order and the reality of her condition in 1944. But the single factor that caused me to question everything was the finding of artificial ultramarine in the blue coatings and what that implied; much use certainly sometime after 1828, which could be a considerable time after, given the time such newly-invented paint would percolate down to be available in West Cork, post 1828.

9.2 A PULLING BOAT?

The other element related to my conviction that the barge was originally built as a pulling boat only, and not rigged to sail, but converted later to do so. For example, the manner in which thwarts No's 2 and 6 were fashioned to support masts. Fashioned pieces of wood fitted to their after edges to make them wide enough amidships to enable the fitting of a mast recess and the fitting of metal mast-bands did not appear to be original (each of these thwarts were the locations for two of the vessel's masts).

The large lodging knees, of Douglas fir, to the same thwarts were ill fitted, one with a filler piece to its side, the opposite knee roughly notched around adjacent frames. Again, presuming the availability at the time of good quality boatbuilding woods, Douglas fir included, the use of knotty Scots pine and birch is unusual in such a vessel. It paints a picture of being very economically built and using some materials that at the time (and particularly in boatbuilding) would be considered inferior.

Figure 10. The conserved *Bantry Boat*, still keeping some of her secrets.

9.3 THWART No 5

But that alone is not the answer to the undersized mast supporting thwarts and their sided pieces. Thwart No 5, for example, is also formed to support a mast, but that thwart was formed wider than the other thwarts to do so – there are no additional widening pieces fitted to it. There is, however, a square recess for a mast in the after edge of the thwart, not a round opening as on the other mast bearing thwarts. The after edge of No 5 thwart was not beaded like the fore and aft edges of all the other thwarts; it was only beaded on the fore edge. Furthermore, each of these thwarts is original and not later replacements. The original fastening sequences of thwarts, knees and sheer planks testify to that. But why form such mast-supporting thwarts differently and indeed why not use the mast at thwart No 6 to carry a pennant or lamp, instead of locating one to do so at thwart No 5? Something wasn't quite right. This would also explain No 5 thwart with its square mast receptacle and broader width. Navy boats all had the means to mount a staff or mast so a flag or lamp could be hoisted, whether they were sailed or not. No 5 thwart in all probability was where a single mast to carry a lamp or pennant was located - in a pulling-only boat. If so it would certainly explain its existence.

Another important point is that thwarts No's 2 and 6 in their single piece form have decorative beading to their forward and after top edges, and so beg the question - why decoratively mould a thwart edge that you intend fitting a widening piece of wood to? In a boat construction context alone it would not be done. Indeed, why not fit thwarts of the correct width in the first place (as with Thwart No 5) if they were to support masts.

10. TRANSOM ALTERATION

10.1 TRANSOM

The transom also shows evidence of alteration related to sailing. A rectangular piece of wood had been fitted to the centre of the inner upper portion of the transom, though placed off-centre towards the port side and through which an iron hinged mast-collar had been fitted centrally to take the mizzen mast. A large square opening had been cut through both fitted piece and transom to provide an opening for a boom to protrude aft of the vessel on the starboard side. Again, this appears to be a later addition to the vessel, not part of its original build. If the vessel had been built to sail the transom would undoubtedly have been formed specifically to support the mast and boom. We then began to think the unthinkable – was she in fact converted to carry masts and sails in Ireland, after her capture?

Furthermore, the iron mast collars in thwarts No's 2 and 6 were found to be fixed with steel countersunk wood screws. Such fastenings were found nowhere else in the vessel. However, while the mast collar from thwart No 5 was missing, its fastenings remain in place – and they are of copper – square copper fixings. Assessing this evidence we became convinced the barge was not originally built to sail.

11. ANALYSIS of THWART'S

11.1 THWART COATINGS

While coatings to all the thwart top surfaces had predominately disappeared, close inspection of both mast-supporting thwarts revealed that the mating surfaces between the thwarts and the widening pieces fitted to them appeared to be coated with blue paint. The widening pieces to both thwarts were removed. The exposed thwart edges showed clearly they were painted, and with several layers that also matched specifically the finish on the other thwart ends, with both blue and red paint still in place on the uncovered thwart edges. It was obvious that the edges of thwarts No's 2 and 6 had been painted a number of times before the widening pieces were fitted, indicating that both thwarts had, for a considerable period, not been fitted with the widening pieces, and in consequence not fitted with mast collars or masts. Furthermore, the mating edges of the removed widening pieces were unpainted, just bare wood.

11.2 THWART COATING ANALYSIS

Paint samples were taken from the exposed after edges of both thwarts. They were found to contain the same pigment found in the blue paint elsewhere on the barge - synthetic ultramarine - which was not invented until 1828 – 22 years after the barge was captured in Bantry Bay. This information, coupled with the manner in which the thwarts were formed to support masts, as was the transom also altered to support a mast, combined to very strongly suggest that the vessel was not originally built to sail, but to be rowed only and to carry a single centrally placed mast, at thwart No 5, to support either a flag or a lamp. And that the sailing aspects were introduced in Ireland.

12. CONCLUSIONS

The term 'the Devil is in the detail' must often be used by conservators when seeking to prove or disprove authenticity or when baffled by conflicting findings. It is a term particularly relevant to the *Bantry Boat.* Given the evidence stated it appears the *Bantry Boat* was originally built as a pulling boat only, with a single mast at No 5 thwart to carry a flag or lamp. That the introductions of seat widening and strengthening to enable sailing were also carried out considerable after 1828 in Ireland. Furthermore, that the alteration to her original colour scheme with the applications of both red and blue coating to her sheer, gunwale and thwarts, both externally and internally, also happened in Ireland.

These conclusions are based on information gleaned from the barge by close examination and consideration. There are many inconsistencies related to the *Bantry Boat* that perhaps require further research of a broader nature than the Conservation Project could allow.

13. ACKNOWLEDGEMENTS

Photographs copyright of National Museum of Ireland.

Jonathan Carr, Head of Maritime and Transport Conservation, National Museums Liverpool. A friend and colleague and major contributor to the successful conservation of the *Bantry Boat.*

Simon Stephens, The National Maritime Museum, Greenwich, for his knowledge and advice related to historic vessels.

Chris Mosely, Ship Model Conservator, National Museums Liverpool, for making me aware of the French Prisoner of War model, and allowing me access to it

14. REFERENCES

1. www.atlanticchallengeinternational.org
2. Siobhan Watts, Conservation Scientist: www.liverpoolmuseums.org.uk
3. Paul M Kerrigan, The French expedition to Bantry Bay, 1796 and the boat from the *Resolue.* Irish Sword Vol. xxi, Military History Society of Ireland, facebook.com/tmhsi
4. The Boats of Men-of-War, Commander W E May. Chatham Publishing ISBN 1 84067 4318
5. Ceremonial Barges on the River Thames, Kenneth Nicholls Palmer, Unicorn Press ISBN 1 0 906290 17 1
6. Ship Models, Their purpose and Development from 1650 to the Present, Brian Lavery & Simon Stephens, National Maritime Museum, Greenwich.
7. Pulling back to Bantry, Classic Boat Magazine, September 1996.

15. AUTHORS BIOGRAPHY

John Kearon is a retired Master Shipwright and Historic Vessel Conservator, traditionally trained in wooden boat-building and lofting with John Tyrrell & Sons, Arklow, Ireland. He spent five years in the Merchant Navy as Chief Petty Officer and Ships Carpenter with both Royal Mail and Shaw Savill lines. He was Head of Department of Maritime, Industrial and Land Transport Conservation with National Museums Liverpool, retiring after 25 years in 2006. From 2007 to 2012 he was with National Museum of Ireland as Project Manager of the *ASGARD* Conservation Project. He is a certified Museums Association Council conservator of wooden vessels and serves on the Advisory Committee of the National Historic Ships Unit. He has also served on the HMS *Victory* Technical Committee. In 2014 he was awarded the Victory Medal by the Society for Nautical Research for outstanding work over time in the Conservation and Preservation of Historic Vessels.

DESIGN EVALUATION AND ALTERATION OF THE DARK HARBOR 17.5: CASE STUDY OF A MODERN REPLICA.

L G Alessio, Southampton Solent University (EMship), UK.
J-B R G Souppez, Southampton Solent University, UK.
A Hage, University of Liège, Belgium.

SUMMARY

While there is still a strong passion and interest for traditional crafts, many of the designs now need to be adapted to comply with the contemporary rules and regulations, and additional modifications are required to meet modern expectations, in areas ranging from comfort to safety; hence the need for modern replicas. In order to ascertain the constraints on modern replicas, a redesign of a Dark Harbor 17.5 will be proposed. First, an initial design evaluation will be performed to assess the characteristic of the original yacht, regarding hydrostatics, hydrodynamics, structural arrangement and stability. Modifications in accordance with regulations and owner's requirements will then be implemented to create a modern replica suited to today's market. Finally, the new design will be compared to the original one, allowing to evaluate the challenges of design modernization and the impact of contemporary requirements on a traditional design.

1. INTRODUCTION

Designed in 1908 by B. B. Crowninshield, the Dark Harbor 17.5 is a classic example of a traditional day sailor. It is the particular lines of historic vessels that make them attractive and elegant and still lead a number of sailors to choose traditional boats. However, true replicas can prove to be unsuitable from a commercial point of view. Indeed, traditional designs were not conceived for contemporary rules and regulations. Plus, what used to be inexistent or a luxury feature, such as an engine on a small sailing vessel, has now become norm. There is, therefore, a call for modern replicas, complying with the relevant regulations and offering a similar standard of comfort as modern boats.

To investigate the challenges resulting from those modern replicas, the redesign of a contemporary Dark Harbor 17.5, currently being built at the IBTC Portsmouth, has been undertaken. The initial design has been assessed, to later allow a comparison with the modern version, with particular emphasis on the issues raised by regulatory compliance and the incorporation of modern comfort elements.

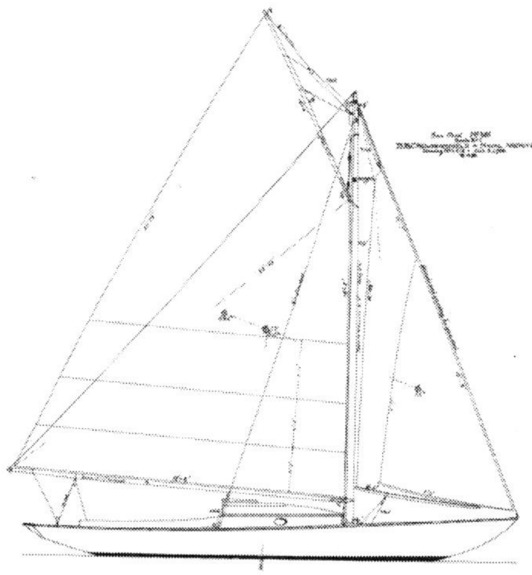

Figure1: Sail plan of the Dark Harbor 17.5 [1].

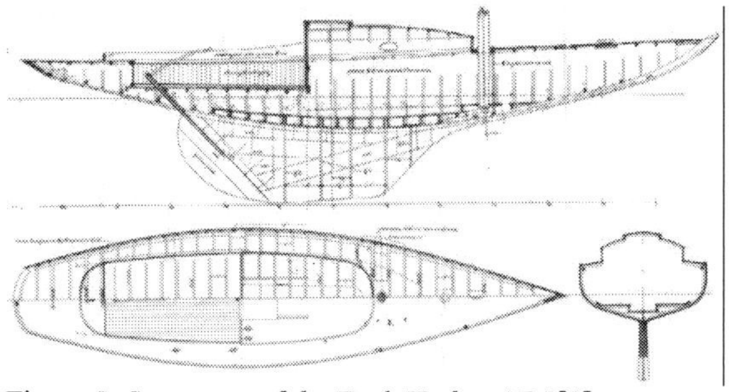

Figure 2: Structures of the Dark Harbor 17.5[1].

2. THE DARK HARBOR 17.5

2.1 ORIGINAL DESIGN

A number of documents relative to the original Dark Harbor 17.5 are still in existence [1] and include: a linesplan, table of offsets and scantlings, as well as a sail and construction plan, respectively presented in Figure 1 and Figure 2.

The main parameters for the Dark Harbor 17.5, derived from the original plans and documentation, are listed in Table 1.

Table 1: Original Dark Harbor 17.5 particulars.

LOA	25' 10''	7.87 m
LWL	17' 6''	5.33 m
BOA	6' 3''	1.90 m
Draft	4' 3''	1.29 m
Displacement	3420 lbs	1551 kg
Sail Area	311 ft^2	28.89 m^2

Finally, the original design philosophy of the craft is captured in the following short historical description of the design [2]:

"In 1908 B.B. Crowninshield was asked to draw up a one-design class of knockabouts to be initially known as the Manchester 17½. The class was to become one of the most popular and long-lived of the knockabouts; about 200 boats were built in Maine, for example, where the name was altered to reflect yacht club affiliation. The most common name for the design is now the Dark Harbor17½, named after the summer colony at Islesboro that once had the largest number of these boats.

One still finds many a Dark Harbor 17½ "knocking about" New England waters. (a dozen or so reside at the Buck's Harbor Yacht Club in Brooksville, where they are still raced on Eggemoggin Reach.) The boats were built well and have lasted well, with cedar planking over oak frames, a lead ballast keel, copper and bronze fastenings, and simple deck construction – canvas-over-cedar-or-pine – to discourage freshwater leaks. Most were built with the self-bailing cockpit as shown on the drawings, although a few were given deep cockpits with seats for more comfort.

While intended primarily for afternoon sailing and racing, these boats have often been used for coastal cruising; the low cabin trunk has space for two transom berths."

Based on all those information, a design assessment of the original craft was performed, as presented in Section 3.

2.2 MODERN DESIGN BRIEF

The Dark Harbor 17.5 was originally designed to sail in America, whereas the modern replica currently being built in the IBTC Portsmouth will be operated from Chichester with a sailing programme focussed on the Solent. Following this first modern build, it is expected that a small commercial production line will be set up.

The design brief, clearly established by the owner, is to build a classic boat with upgrades from the original to make the boat more comfortable and practical; the main requirements being:

- Increase the operational area, by making the keel shallower, and therefore be able to expand the sailing programme of the boat. This is also linked to the intention of making the boat trailerable.
- Modify the hull planking from carvel to strip planking with mahogany veneers. This is aimed at achieving a stronger and more modern construction while minimising the maintenance for the owner.
- Add an electric inboard engine to make the boat safer and practical. Indeed, very few owners would consider sailing into harbour nowadays.
- Add a portable toilet to enhance the level of comfort for the owner and his guests; this also extends the sailing programme and enables spend a night onboard.
- Add an anchor locker in a watertight compartment at the front of the boat.
- Change the material of the mast, boom, gaff and fore boom from wood to carbon fibre, making the boat lighter.

The impact of these requirements is significant, not only for the redesign but also regarding the type of replica that will be achieved.

2.3 MODERN REPLICAS

There is great importance in being able to state whether a vessel can be considered a replica or a reconstruction. A number of definitions have been laid out by the National Historic Ships UK [3]. For instance, the process of replication is given as: *"Replication means starting from scratch to build a copy of a vessel and can be defined at various levels of detail and accuracy".*

A true replica would be characterised as an exact reproduction of the original vessel in every detail, except some minor details such as the fender ropes.

Since the Dark Harbor will deviate from the original, the definition of an operational replica seems better suited, as it allows for some modifications to meet the operational needs, such as the safety equipment and some internal changes for specific functions.

However, the major changes to the keel design and radically different construction method make the modern Dark Harbor a representation. Indeed, the representation is defined as a vessel which was based on a particular craft but had its appearance modified, considering the overall impression but not the accuracy.

The vessel is to undergo some major changes, dictated by the owner in order to meet the contemporary standard of comfort expected, but the vessel is to retain the spirit of the original Dark Harbor, thus making it a representation project.

3. DESIGN ASSESSMENT

3.1 3D MODELLING

The hull surface was modelled with a computer aided design (CAD) software, following the instructions of the linesplan and the table of offsets, making the hull as close to the original design as possible. As the original design dates from 1908, there is some uncertainty regarding the accuracy of the lines and the measurements. Originally, lofting the boat full size would have ensured fairness; in this case, too, some corrections were necessary to make the hull lines smoother, leading to the model presented in Figure 3.

Figure 3: 3D model of the original Dark Harbor 17.5.

To check the accuracy of the designed model, the hydrostatics were calculated based on the linesplan using Simpson's rule and then compared with the hydrostatic analysis of the 3D model. This comparison is shown in Table 2.

Table 2: Modelling comparison

Particular	3D Model	Simpson's Rule	Error
LOA (m)	7.92	7.92	0%
Lwl (m)	5.34	5.33	0.13%
BOA (m)	1.91	1.91	0%
Bwl (m)	1.84	1.82	0.85%
Tk (m)	1.28	1.28	0.05%
Displ. (kg)	1555	1542	0.86%
WSA (m^2)	12.64	12.06	4.62%
WPA (m^2)	6.61	6.78	-2.60%
Cp	0.47	0.50	-5.56%
Cb	0.12	0.12	0.13%
LCB % from AP	45.41	45.18	0.49%
LCF % from AP	44.28	45.97	-3.81%

To use Simpson's rule the waterline was divided in ten sections, and this may have affected the accuracy of some calculations. To increase the accuracy, a greater number of sections may be required. However, the relatively small errors demonstrate a faithful representation of the original design.

A detailed weight estimation was also performed based on the original scantlings and allowed to validate the displacement of 1556 kg, with a vertical centre of gravity (VGC) 272mm below the design waterline, and a longitudinal centre of gravity (LCG) 210mm aft of amidships.

3.2 STABILITY

In order to evaluate the original design, its stability will be tested against modern regulation. This is critical to ensure that the new boat, despite the shallower keel, will have a stability at least equivalent to the original.

All the recreational crafts placed on the market within the European Economic Area (EEA) are to comply with Recreational Craft Directive (RCD) [4]. Part of the requirements to make the boat available on the market is to follow the guidelines of the ISO 12217 for stability to demonstrate compliance of the craft with one of the four design categories. In this case, the vessel being intended for coastal cruising, it will be designed for category C, inland, allowing sailing in wind speeds up to Beaufort 6, and significant wave heights up to 2m.

As a vessel of hull length greater than 6m, the vessel is to comply with the ISO 12217-2 [5]. This involves consideration for a range of parameters, such the downflooding angle (DFA), righting moment (RM), the angle of vanish stability (AVS) and wind stiffness (WS). These values were assessed separately and then compared in Section 4.8 with those obtained after all changes were implemented.

3.3 STRUCTURE

The original structure of the boat was made of ⅞ inches (22.22 mm) square oak frames, evenly spaced every 8 inches, and steamed onto the ¾ inches (19.05 mm) cedar carvel planks, fastened with copper rivets. An analysis of the structure against present regulation, namely the ISO 12215-5 [6], was realised, highlighting that a factor of safety of over 5 was applied for the frame spacing.

It is however critical to mention that this analysis was purely indicative and that the results cannot be considered fully accurate. Indeed, the ISO 12215-5 idealises the panels and stiffeners as built-in beams. Although unknown, the end fixity of a plank riveted to timbers cannot be 100%, thus introducing a significant uncertainty on the results. Furthermore, the ISO 12215-5 is not meant to be applied to traditional wooden construction. While only a limited structural assessment was possible for the original vessel, the new structure designed and built using modern construction techniques fall within the scope of the standard and the assumptions made in the underpinning theory will be valid, thus providing a much more relevant analysis.

3.4 PERFORMANCE

An important aspect of the performance evaluation was the resistance model, based on the Delft Systematic Yacht Hull Series (DSYHS) [7]. The total hydrodynamic resistance of the boat is given by the sum of the resistances of the hull and the appendages and will be compared to the new design in Section 4.9.

4. REDESIGN

4.1 MOTIVATIONS

The design was modified according to the client's requirements presented in Section 2.2, with the objective of retaining a classic sailboat, with however some modernization to improve the level of comfort and comply with the rules and regulations.

The addition of a propulsion system, toilet, modern electronic equipment, etc. will result in an increase in weight. This is the motivation behind a modern construction and carbon rig: trying to save weight to balance the addition of the equipment onboard and keep the vessel on its original waterline.

4.2 KEEL DESIGN

The new keel aims to keep the performance and stability of the original boat. As a result, a number of parameters were kept the same, such as the centre of lateral resistance (CLR), the planform area and the sweep angle. For rudder, the shape had to be changed as a consequence of the shallower keel, but the area was conserved, thus keeping the original manoeuvrability of the boat.

The challenge on the keel design was to reduce the draft with the lesser impact on the VCG as possible. Indeed, the weight added in the hull and a shallower keel, imply a higher VCG, and therefore a reduction in stability which should be avoided.

As a consequence of reducing the draft, the chord length will be increased to conserve a similar planform area. The longitudinal position of the keel must also be modified in order for the CLR to be kept in the same location. Finally, the longer root chord leads to a longer lead section at the bottom, thus allowing to reduce its height for a given volume and lower the VCG for the given span of the keel.

Different types of keel were designed and the stability was checked for all to ensure compliance with the ISO 12217-2. The final design was selected by the owner based on aesthetical considerations, as well as the ability to save weight. Indeed, to retain a given righting moment, either a larger mass must be provided acting at a smaller lever, or vice versa. In this case, three options were proposed:

- A 1.00m deep keel with a 750kg ballast (50kg heavier than the original).
- A 1.05m deep keel with a 700kg ballast (same as the original).
- A 1.10m deep keel with a 630kg ballast (70kg lighter than the original).

The latter option, depicted in Figure 4, was adopted, thus giving a 7.5 inches draft reduction compared to the original, and now allowing the structural arrangement to be tackled.

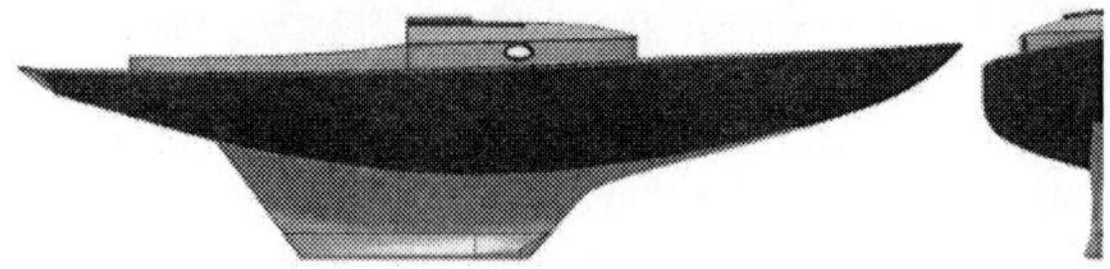

Figure 4: New shallower keel.

4.3 STRUCTURE

A modern wooden hull construction method has been selected, composed of 12mm thick western red cedar strip planking, then covered by two layers of 2.4mm mahogany veneers, placed at +/- 45 degrees; hence a total hull thickness of 16.8mm.

As the pressures acting on the hull decrease further aft, the frame spacing took advantage of this. Consequently, the 350mm spacing applied forward of station 7.5 is increased to 450mm aft. One of the originalities of construction is to retain square stream-bent oak frames, to minimise the labour and cost associated with laminated frames.

The new structure was checked against the ISO 12215-5 requirements for category C, thus ensuring compliance with the regulation.

4.4 PROPULSION SYSTEM

To be more manageable in marinas, add safety in emergency situations and be able to operate against the strong tidal currents found in the Solent, an engine must be fitted. As the vessel is aimed at inshore operation, only a limited range is required. This, associated with the will to achieve a sustainable design and ensure quiet motoring, lead to the choice of a fully electric propulsion.

Based on the resistance of the new Dark Harbor 17.5, the power required to achieve a given speed was ascertained, and is shown in Figure 5. Note that an efficiency of 0.69 was considered for the chosen FeatherStream propeller [8], established based on the Wageningen B-Series Propellers method [9].

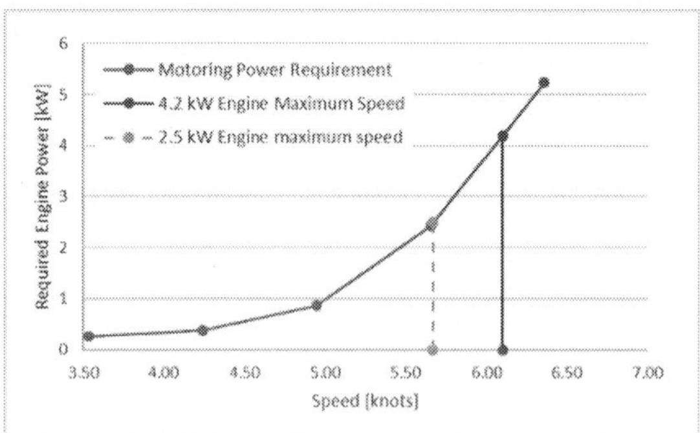

Figure 5: New design power curve.

The considered Bellmarine electric engine [10] being available in 2.5kW or 4.2kW versions would result in top speeds of 5.6 and 6.1 knots respectively (neglecting the air drag and added resistance due to waves at this stage). Eventually, the 4.2kW engine was chosen.

This then allowed to assess the number of batteries needed and associated range. With a capacity of 138 Ah, each Valence battery [11] would provide a range of 8.2 nautical miles at maximum speed for a weight of 19.5 kg. A total of 4 batteries were selected, thus providing a range of 32.7 nautical miles when motoring at full speed. However, the range is significantly increased at lower speeds, as presented in Figure 6.

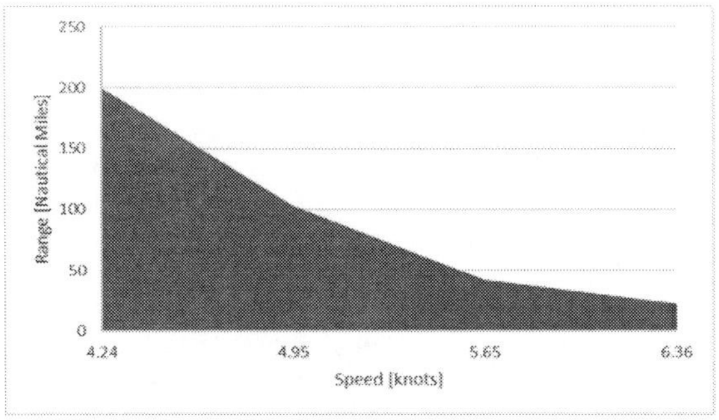

Figure 6: Motoring range for different speeds.

It is to be noted that this range does not include considerations for the use of navigation equipment and navigation lights at night, but it considers the recommended maximum discharge of 80% of the battery capacity to maximise its lifespan.

4.5 AESTHETICS AND EQUIPMENT

To make the boat more visually attractive, the option of teak decking was adopted. Indeed, with a low freeboard, a teak deck enhances the aesthetical appeal of the design. This represents an increase in cost, but also weight; 20kg in this case.

An anchor locker was added at the bow of the boat in a watertight compartment. This fit with the operation philosophy of the boat, and enables not compromising the appearance of the vessel.

A portable chemical toilet was added, chosen for its minimum volume as the available space in the boat is very restricted. This addition makes the sailing experience more comfortable, especially when sailing with guests.

As per the engine location, the installation of the toilet highlighted a major challenge of modern replicas: the extremely limited space available to implement all the required modifications to meet the standard of comfort expected nowadays. Indeed, no spare volume was allocated to fit an engine or a toilet in the original design, making the modernisation particularly complex.

4.6 CARBON FIBRE RIGGING

The original rigging was composed of wooden spares, namely the mast, boom, gaff, and foresail boom, as depicted earlier in Figure 1. Having raised the centre of gravity because of the shallower keel and additional equipment installed, moving to carbon fibre spars appears particularly attractive, as saving weight very high will significantly lower the overall VCG of the boat.

Firstly, the foresail was modified, and the use of a furling system implies that no foresail boom will be required.

Then, the original keel-stepped mast (coming through the deck and resting on the backbone) was replaced by a deck-stepped mast, supported by a mast pillar inside the hull. From a structural point of view, this means that a stronger and therefore heavier mast section will be required, due to the difference in end-fixity between a keel-stepped and a deck-stepped masts. This decision is however dictated by the necessity for the owner to be able to take the mast down; also ensuring a better watertight integrity of the deck, critical on a wooden boat.

In order to evaluate the weight savings of a carbon mast, boom and gaff, a rig design based on the Nordic Boat Standard [12] was undertaken, and allowed to calculate the reduction in the weight of the rig thanks to the use of carbon fibre. Furthermore, the sails will be replaced by a high-quality polyester sailcloth, Dacron, which has greater resistance and durability compared to the canvas. Other sources of weight savings include new shrouds and stays as well as deck fittings and ropes.

Overall, the new rigging will allow saving an estimated total of 23kg, now allowing to perform a weights and centres calculations for the new boat.

4.7 WEIGHTS AND CENTRES

While the weight change is critical in the design process, so is the location of the centre of gravity. Hence the need to update those parameters for the new vessel. The balance between the weight gained and the weight saved compared to the original is presented in Table 3.

Table 3: Weight changes

Weight change	Mass (Kg)
Gained	143
Saved	
Structure	25
Rigging	23
Ballast	70
Total:	**25**

Despite the addition of several new elements to the boat to fit with contemporary standards, which includes engine, batteries and a toilet, savings in other locations brought the overall weight to 1595kg, i.e. a 40kg increase, with no alteration to the location of the VCG. This allows the vessel to float on its original waterline, and not to impact the original stability.

4.8 STABILITY

The results of the stability assessment are summarized in Table 4, and compared to the requirement for category C under the ISO 12217-2.

Table 4: Stability Comparison.

Parameter	Original	New	Requirement
DFA (°)	58.1	56.3	>35
AVS (°)	137.7	137.4	>90
WS (°)	38.0	38.5	<45
Max. Gz (m)	0.50	0.50	-
Max. Gz (°)	79.0	78.0	-

Despite the shallower draft and heavy modifications, the overall hydrostatics and stability of the boat were retained, ensuring the original qualities of the vessel are conserved, while meeting the regulatory requirements.

4.9 IMPACT ON PERFORMANCE

With only a negligible increase in displacement and a constant wetted surface area despite the shallower draft, as the keel area was kept constant, the overall resistance increase is 2.7%.

Furthermore, as there is no significant change in the stability and the sail plan will remain the same, it can therefore be concluded that the speed of the boat will not suffer any significant alterations, as validated by a velocity prediction programme.

As a result, the general appearance and performance of the original boat were conserved.

4.10 CONCLUSIONS

The modifications implemented to the Dark Harbor 17.5 design made the vessel more in line with today's customer expectations. The addition of an engine, shallower keel and toilet offer greater comfort, while the modern construction technique and carbon fibre rig provide an attractive commercial argument, coupled with a minimum maintenance required. Furthermore, compliance with the RCD category C was demonstrated. All of these did not actually impact the behaviour and performance of the boat, and the aesthetical qualities of the original have been conserved.

5. CONCLUSIONS

Thanks to the availability of historical designs, modern replicas can now be built with great accuracy, with however contemporary requirements for a more comfortable and safer design, complying with the appropriate rules and regulations.

The design of a modern Dark Harbor 17.5 replica, or rather representation, was undertaken to significantly modify the engineering of the vessel, while retaining the original aesthetical attraction inherent to classical designs. A shallower keel allowed to increase the area of operation and make the boat trailerable. The electric propulsion system allows for a more practical use of the vessel, while acknowledging the modern environmental concerns. The addition of comfort such as a toilet also makes the vessel more commercially attractive. Furthermore, the modern wooden construction, with strip planking covered with veneers makes for a strong and durable hull, requiring minimum maintenance; while the carbon rigging allows for weight savings and easier handling of the craft. Finally, the design was proven to comply with the RCD and relevant ISO standards for category C.

The design assessment of the original craft allowed to ensure the modern representation would perform and behave the same; this was achieved with a virtually identical stability and resistance, an untouched sail plan, and a keel and rudder arrangement conserving the characteristics of the original.

However, significant challenges were highlighted in the design process. The additional components to be retrofitted to the original design are of significant volume of mass, particularly on a vessel as small as the Dark Harbor 17.5. Physically positioning those such as the engine, for instance, can be particularly challenging. Furthermore, it is very easy to make the boat heavier, and new means of saving weight in other areas must be found, for instance looking at a lighter structure, more efficient use of the ballast, or modern carbon fibre rig. Finally, modern regulations, particularly for structure, are not suited to analyse traditional construction methods such as carvel.

Nevertheless, those challenges can be overcome to provide modern replicas and representations, commercially attractive and compliant with the present regulatory framework, while capturing the essence of the original design and therefore supporting the conservation of historic crafts and classic designs.

6. ACKNOWLEDGEMENTS

The principal author would like express special thankfulness, warmth and appreciation to the persons and organisations below for their support in the realisation of this project.

The professors of the EMShip Master programme for their vital support and assistance.

The professors of Southampton Solent University for all their patience in sharing their knowledge and expertise.

The owner of the Dark Harbor replica and the IBTC Portsmouth, where the built is under way, for the opportunity to be involved with the redesign and the fascinating discussions.

7. REFERENCES

1. CROWNINSHIELD, B. B., 'Dark Harbor 17½', *Dark Harbor Design Plan,* 1908.

2. WOODEN BOATS, '25'10" Sloop, Dark Harbour 17½', *Wooden Boats*, undated.

3. HEIGHTON, M., 'Replicating Historic Vessels,' *National Historic Ships - United Kingdom*, 2012.

4. EUROPEAN PARLIAMENT, 'Directive 2013/53/EU', *Official Journal of the European Union*, 2013.

5. ISO, 'BS EN ISO 12217-2:2015 - Small Craft – Stability and buoyancy assessment and categorization. Part 2: Sailing boats of hull length greater or equal to 6m', *International Organisation for Standardization*, 2015

6. ISO, 'BS EN ISO 12215-5:2008+A1:2014 - Small Craft – Hull Construction and Scantlings. Part 5: Design Pressures for Monohulls, Design Stresses, Scantlings Determination', *International Organisation for Standardization*, 2014

7. KEUNING, J. A. and SONNENBERG, U. B., 'Approximation of the Hydrodynamic Forces on a Sailing Yacht based on the Delft Systematic Yacht Hull Series', *The International HISWA Symposium on Yacht Design and Yacht Construction*, 1998.

8. DARGLOW ENGINEERING, 'FeatherStream Propellers data sheet', *Available online at: http://www.darglow.co.uk/*, 2015.

9. BERNITSAS, M. M., RAY, D. and KINLEY, P., 'K_T, K_Q and Efficiency for the Wageningen B-Series Propellers', *The University of Michigan*, 1981.

10. ID TECHNOLOGY BV (BELLMARINE), 'Propeller Shaft Propulsion – User Manual', *available online at: www.bellmarine.nl,* 2013.

11. VALENCE, 'U-Charge® XP Modules', *available online at: https://www.valence.com,* 2016

12. LARSSON, L. and ELIASSON, R., 'Principles of yacht design', *Adlard Coles Nautical, London*, 2007.

8. AUTHORS BIOGRAPHY

Luis Gabriel Alessio dos Santos is a Mechanical Engineering graduate, currently undertaking the Erasmus Mundus programme in Advanced Ship Design and Maritime Engineering (EMship).

Jean-Baptiste R. G. Souppez holds the current position of Lecturer in Yacht Design and Composite Engineering at Southampton Solent University. He is responsible for delivering a number of units of the BEng (Hons) Yacht and Powercraft Design and BEng (Hons) Yacht Design and Production, and is a member of the ISO working group 18 (ISO 12215).

André Hage holds the current position of Professor at the University of Liège. He is responsible for teaching ship design as part of the EMship Master programme, and is the director of the towing tank.

A DUTCH BOEIJER YACHT BROUGHT TO LIFE

A.P. de Vos, Traditional Boatyard "Klaas Hennepoel", The Netherlands

SUMMARY

In Warmond, in the Netherlands, a traditional boatyard was reconstructed, using authentic materials and constructions. At this boatyard the 19th century boeier yacht 'Vivo' is being restored. Next to the restoration of this carvel build boeier, a project was started concerning an even earlier type, the clinker built boeijer yachts. Although a considerable number of carvel boeiers is still navigating Dutch lakes and canals, only ship models, pictures and drawings of their clinker built ancestors remain. A research project aims at deepening the knowledge around this type of yacht in general and especially on the boeijer that was presented by the Dutch king Willem I to the Russian emperor Alexander I in 1816 on the occasion of the marriage of crown prince Willem II to the sister of the emperor, Anna Paulowna.

The project combines historical research, e.g. in the archives of the Navy shipyard which built the boeijer for the emperor, with technical research of a number of admiralty models of the boeijer. Both traditional methods and modern techniques are used to investigate the lines and construction of the models. A purpose built measuring tool was used to transfer the shape of the stations onto paper. Furthermore a 3D laser scan was made and a CT-scan. An independent source of information was found in the drawings that were made from the full size boeijer shortly before it was demolished in Leningrad in 1954.

Investigation of the hydrostatic and hydrodynamic properties is the next step. A tempting follow-up is the reconstruction of the boeijer.

1. INTRODUCTION

In the Netherlands yachting started as early as the 17th century. Most types of yachts of 4 centuries ago, like the 'speeljacht', 'hekjacht' or 'paviljoenjacht' do no longer exist. However, there still remains a substantial fleet of Dutch boeier yachts in use. These wooden boeiers strongly resemble their 17th century predecessors. Most boeiers that have survived were built in the second half of the 19th century. The boeiers that are still in existence are all carvel built, however there used to be clinker built boeijers as well (note the difference in spelling between carvel built boeiers and clinker built boeijers, this will be elaborated later). It was assumed that the last remaining clinker built boeijer was "Stavo", broken up in 1923. At the traditional boatyard "Klaas Hennepoel" a project is underway to do research on the clicker built boeijer yachts which may eventually even lead to the reconstruction of this extinct type of boat.

Construction of the historical boatyard started as a private initiative. Eventually the project was supported by the Provincie Zuid-Holland, by the municipalities of Teylingen and Oegstgeest and by the EU Rural Development Programme. Objective of the boatyard is to preserve local traditional boats, materials and boatbuilding skills from the 18th till early 20th century. The boatyard sheds were built using authentic constructions and materials: Original 'Rhine-size' bricks were laid using lime mortar, the wooden structure was joined together using wooden nails to fasten the mortise and tenant joints, recycled traditional 'oude holle' roof tiles were used, weather boarded 'gepotdekseld' facades, etc. After many years of preparation, construction work started in 2010. The slipway and timber shed were completed in 2012 and the boatbuilding shed was completed in 2014. Oktober 2014 the 19th century boeier "Vivo" was transported to the boatyard in order to continue its restoration (fig. 1).

Figure 1: Transport of the 19th century boeier "Vivo" to the traditional boatyard "Klaas Hennepoel".

The traditional boatyard is located in Warmond along the Haarlemmertrekvaart, the tow barge canal between the cities of Leiden and Haarlem, which was established in 1657. The city of Leiden owned a clinker built boeijer during the 18th and 19th century. This so called 'Stadsboeijer' or 'Vroonboeijer' was used as a means of transportation for the city (stad) governors and to inspect the fishing rights of the city on the 'Vroonwateren'.

In 2014 the Maritime Museum in Amsterdam granted the Mr. P. Rogaar Fellowship to do research into the clinker built boeijer yachts. This project concerned both historical as well as technical aspects of clinker built boeijers. The project focussed on the boeijer that was commissioned in 1814 by the Dutch king Willem I to be built at the Navy shipyard in Amsterdam. This boeijer was presented to the Russian emperor Alexander I in 1816 on the occasion of the marriage between crown prince Willem II and Anna Paulowna, sister of the emperor.

The key research topic was to unveil the story of the ship model of the Boeijer for the Russian emperor in the collection of the Amsterdam Maritime Museum (fig. 2). Four issues were addressed:
- What are the characteristics of the Boeijer?
- Why did the Dutch king present a boeijer to the Russian emperor?
- What can be found about the history of the emperor's Boeijer?
- What can be learned from the ship models of the Boeijer?

Figure 2: Model of the Boeijer for the Russian emperor. (collection Amsterdam Maritime Museum, inv.no. S.0897)

2. ANALYSIS OF THE TYPE: BOEIER VERSUS BOEIJER

A study of contemporary sources, ship models, pictures and documents, was made to analyse the characteristics of clinker built boeijers.

In the twentieth century it became a habit to distinguish between boeiers (carvel build) and clinker built boeiers, in Dutch 'overnaadse boeiers'. However, in preceding ages the term 'overnaadse boeier' was not used. In fact, in the 18th century the carvel built boeier type of yachts was designated as 'kopjacht', while the clinker built boeiers were simply termed 'boeijer' or 'boeyer'. A 'kopjacht' shows a strong resemblance to the boeiers that are still in existence: carvel planning, sturdy round bows, upright sternpost and stem, narrow rudder, strong rubbing strake (fig. 3). However, the stem and sternpost of a boeijer are less vertically orientated, the rudder is wide with a triangular shape and rubbing strakes are lacking. The planking of a boeijer is overlapping, but only to some extent. Studying contemporary models learned that in most cases the two upper stakes are overlapping. The lower stakes however are only overlapping above the waterline, thus in longitudinal direction there is a transition from overlapping seams to carvel seams. The bottom strakes are entirely carvel (fig. 4).

Figure 3: Friso, one of the 19th century carvel build boeiers still sailing, mainly in Friesland

Figure 4: Planking of the model of the Boeijer for the Russian emperor (collection Amsterdam Maritime Museum S.0897)

Surprisingly in some cases it occurs that all characteristics of a boeijer (i.e. the type that nowadays would be called clinker built boeijer) are present, however the planking is entirely carvel. Examples are two similar models, one in the collection of the Amsterdam Maritime Museum and one in the collection of the Rijksmuseum as well as two ice boats, one in the collection of the Zuiderzee Museum and the other one in the collection of the Streekmuseum Krimpenerwaard.

It can be concluded that the type of planking is not the main discerning characteristic between 'boeiers' and 'boeijers'. Combined characteristics, i.e. shape of the hull, rudder, presence or absence of a rubbing strake and various details do define the difference between the two types.

3. MOTIV

It is intriguing why the Dutch king presented a relatively small and modest yacht to the Russian emperor. A clue was found in a painting by Abraham Stolk (fig. 5). This painting shows a mock sea battle that was performed in Amsterdam in 1697 to honour Peter the Great. Peter the Great can be seen aboard a fine admiralty yacht, however in the middle of the scene, a modest boeijer yacht is sailing, flying a flag with close resemblance to the Russian flag. An explanation is given in the book by Scheltema [1], published in 1814, about the visit of Peter the Great to Holland in 1697. Scheltema based his book on the information from the dairy of a merchant, Noomen, who observed the visit of Peter the Great in 1697. Peter travelled to the centre of Dutch ship building, the Zaanstreek. His intention was to learn the ship building skills by working incognito at the shipyards. However, due to his non Dutch looks and great length, he was recognised as being the Russian emperor by local craftsmen that had worked in St. Petersburg. Occasionally crowds would gather and Peter would choose to leave. As a means of transportation he bought a boeijer for 450 guilders.

When in 1814 emperor Alexander I visited the Netherlands, almost 120 years later, Scheltema presented his book about the visit of Peter the Great to Alexander I. So it is plausible that the boeijer that was presented by Willem I to Alexander I was meant to be a replica of the boeijer owned by Peter the Great. In the correspondence between Willem I and Alexander I it becomes clear the boeijer was a wish of the emperor himself.

Figure 5: Mock sea battle to honour Peter the Great, September 1st 1697, painting by Arbraham Storck (collection Amsterdam Maritime Museum, inv.no. A.0280).

4. BUILDING THE EMPERORS BOEIJER

On August 1st 1814 the chief engineer of the Dutch Navy Shipyards, P. Glavimans, ordered the chief shipwright of the Amsterdam Navy shipyard, Pieter Schuyt jr., to build a boeijer for the Russian emperor. In his letter Glavimans refers to the accompanying design drawings. Alas these drawings have not been found. Glavimans explicitly states that, in deviation from the drawings, the cabin top should be made removable. This is remarkable as the boeiers that are still in existence never have a removable cabin top. Contemporary models of boeijers do have a removable cabin top, however, as these models all have intricately detailed interiors, it is only natural that the cabin tops can be removed from the models in order to be able to see these interiors. The note by Glavimans sheds a different light, i.e. a removable cabin top may have been an actual option. This is further confirmed in the correspondence in 1815 between Pieter Schuyt jr. and the Navy Ministry. Schuyt pleads for a fixed cabin top as a removable cabin top would not result in a stable construction. Inevitable leakage would spoil the interior. However, the Ministry orders Schuyt to make a removable cabin top, stating that if the emperor prefers a fixed cabin top, this can be arranged after arrival in Russia.

The correspondence between Schuyt and the Ministry addresses more interesting details. Schuyt consults the Ministry about the interior, whether a fireplace should be installed, whether there should be fixed benches or easy chairs, etc. Also Schuyt inquires about the required finish of the exterior and describes the usual colour scheme. The Ministry replies that the boeijer should be left just in primer.

Finally in 1816 Schuyt receives orders to make provisions on deck of the merchant ship 'Kroonprins der Nederlanden' in order to stow the boeijer as a deck load. This is surprising as in 1935 J.C.M. Warnsink writes in his book [2] on the Amsterdam 'Kweekschool voor de Zeevaart' that two students, Salomon Dedel and Willem van Zuylen, sailed the boeijer to St. Peterburg under supervision of navy captain Lantsheer. Warnsink reports that along the way the pupils performed excellent manoeuvres, to great satisfaction of their commanding officer. This would not be possible if the boeijer was transported to St. Petersburg as a deck load. Confirmation is found in the newspapers of July 1816, which state that in June 1816 a merchant ship had arrived in Kroonstad carrying the Boeijer for the emperor.

The reports of captain Lantsheer, who was the personal aid to the king, give a wonderful story of is adventures along the journey and at the imperial court in St. Petersburg.

A Russian publication [3] reports that just three years after the arrival of the Boeijer, in 1819 a copy was built by shipwright Kurepanov at the exact same dimensions: 9,2 x 2,9 x 1,1 m. This ship was simply addressed as 'Boeijer no. 2'. This publication also reports that the Boeijer no. 1 was dismantled in 1890. However, in 1954 the remaining hull of the Boeijer (unclear whether it is no. 1 or 2), was demolished in Leningrad. Before scrapping, pictures were taken, measurements were taken and drawings were made. Even a few decorated parts were preserved and are still part of the collection of the St. Petersburg Naval Museum.

5. THE SHIPMODELS

A major source of information on the Boeijer for the Russian emperor are a number of ship models. Almost a century ago, the descendents of Pieter Schuyt jr., chief shipwright at the Amsterdam Navy shipyard, donated a very detailed model to the Amsterdam Maritime Museum (further on to be addressed as the 'Schuyt model'). This model was the focus of the Fellowship study. Also in the 'Marinemodellenkamer' (admiralty model chamber) that is part of the Rijksmuseum collection, a model is preserved of which the records state that it is the Boeijer for the Russian emperor (to be addressed as 'Navy model')(fig. 6). Furthermore a model which shows great resemblance to the aforementioned models turned up in a private collection. Further on this model will be addressed as 'model 1819' as an ornament at the lower front side of the mast, the 'mastbord', is engraved '1819'. The models differ in dimensions:

Schuyt model	58,8 cm
Navy model	46,3 cm
Model 1819	41,5 cm

Figure 6: Model of the emperor's boeijer part of the ' Marinemodellenkamer' (Collection Rijksmuseum inv.no. NG-MC-675).

6. MEASUREMENTS

Both traditional methods and modern techniques were used to investigate the lines and construction of the models:
- The interior of the models was observed by means of an endoscope,
- A purpose built measuring tool was used to transfer the shape of the stations onto paper,
- A 3D laser scan was made,
- A CT-scan was made.

6.1 ENDOSCOPIC OBSERVATIONS

A main purpose of the visual inspection of the model interiors by means of an endoscope was to ascertain whether the models are signed or dated in hidden places. Especially the forecastle could not be directly inspected. Alas no signatures, dates or other relevant notes were observed. The only written text was observed at the bottom side of the sheer clamps of the Schuyt model. The port sheer clamp was marked 'BB', which is likely to be short for the Dutch word 'bakboord', meaning port. Similarly the starboard sheer clamp was marked 'SB', short for 'stuurboord'.

However, an interesting observation was that the panelling in forecastle was very similar for all models (fig. 7). This suggest that the models were made from the same design drawings.

Figure 7: Endoscopic view of the panelling in the forecastle of the Navy model.

6.2 HAND MEASUREMENTS

After a survey, it was concluded that in the Dutch musea there was no tool available to conventionally measure the lines of a ship model. Therefore a purpose built measurement system was designed. Starting from a complicated design, consisting of mechanical gauges to establish the position of a probe in two directions along a station, the concept was simplified. The idea was that each step of measurement and transfer introduces errors. So a system was developed that transfers the shape of the model directly onto paper (fig. 8): The model is placed level on top of a sheet of paper on which a grid of stations and buttocks was prepared. The probe consists of a rectangular wooden block with a number of holes. A wooden probe is attached to the block by means of a bolt and wing-nut. The side of the block is aligned with a station line and the front of the block is aligned with the outer longitudinal line. The end of the probe is brought to the model hull and fixed in its position by tightening the wing-nut. Then the block is transferred to a drawing board and aligned with the sides of the board on which a sheet graph paper is fixed. The point of the probe is marked on the sheet of paper. Step by step each corner of the planks is thus transferred onto paper, without measurements having to be taken. A transparent sheet of polyester film is aligned on top of the graphs to draw the stations. From there the side view and plan view are derived (fig. 9) The transparent polyester film enables the comparison of the lines of different models by putting them on top of each other.

In order to evaluate the method, the procedure was repeated for some stations to check whether the method results reproducible graphs. The results showed to be identical for repeated measurements.

Not only the stations were recorded with this method, but also the shape of the bow and rear were recorded by measuring along the buttocks. As the bows show a very strong curvature, this is essential to truly capture the shape of the hull.

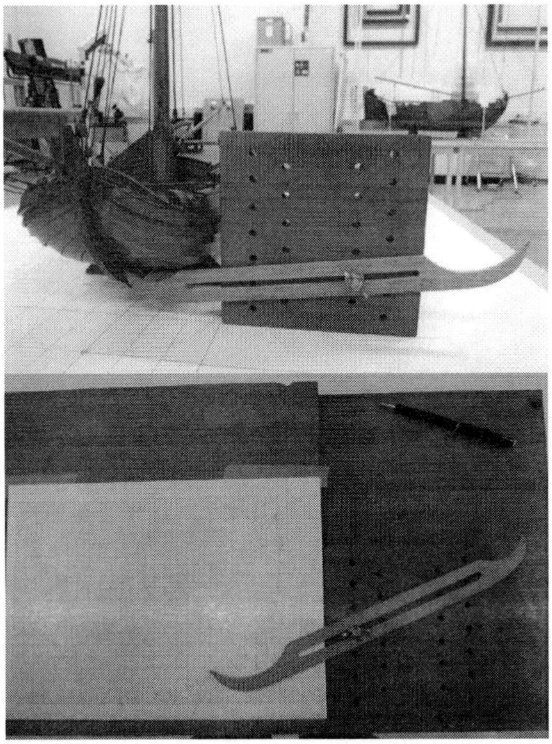

Figure 8: Probe aligned with the stations and transfer of the probe position onto the drawing board.

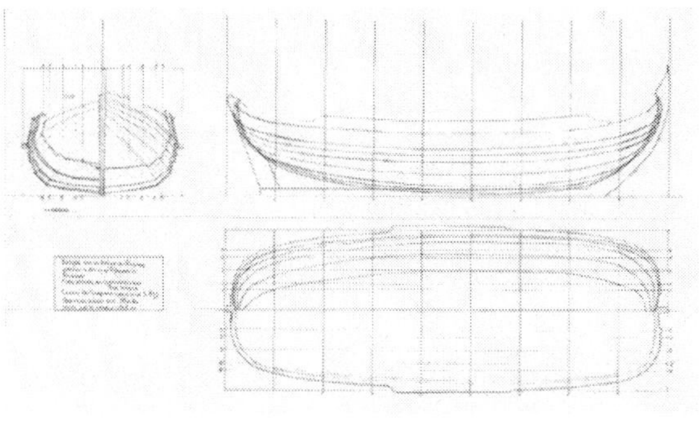

Figure 9: Lines plan of the Schuyt model.

This method was applied to three models: The Schuyt model, the Navy model and a beautiful half model of a clinker built boeijer also being part of 'Marinemodellenkamer' (fig. 10). No information about the provenance of this 58.8 cm long half model was available.

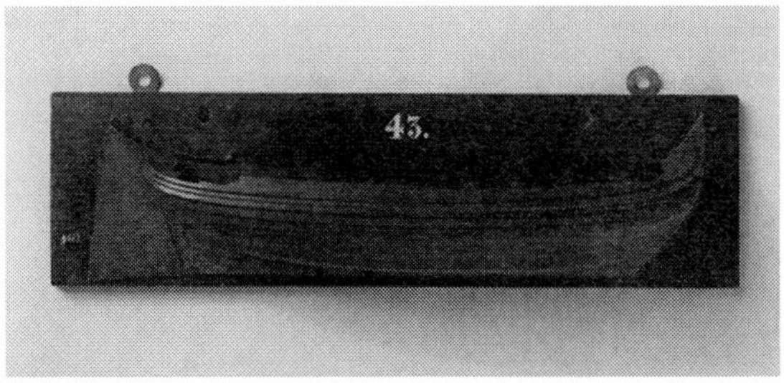

Figure 10: Half model of a clinker built boeijer, part of the 'Marinemodellenkamer' (collection Rijksmuseum inv. no. NG-NM-9417)

The lines plan of the Navy model was copied to the same length as the other two models (both 58,8 cm) to enable comparison. This showed that, although there are differences in details as well as in the shape of the bows, the stations of all three models are close to identical. This suggests that the models are built from the same lines plan. As the original lines plan has not been found, it can be stated that the present approach has resulted a reliable reconstruction of the original lines plan. The measurement method is an example of how a simple but smart approach can give a valuable result. This fits well into the general focus of the traditional boatyard to demonstrate that traditional hand tools, authentic materials and constructions can be efficient and yield good results.

6.3 3 D LASER SCAN

A 3D laser scan of the Schuyt model was made by means of a Faro Edge Scan Arm equipped with a HD Laser Probe (fig. 11). The scan results a very large point cloud file. Geomagic was used to wrap a polygon mesh around the point cloud model. The scan shows some gaps in the data where the overlapping planks shadowed the edge of the lower plank. Next step is to further process the data into a lines plan and construction drawing.

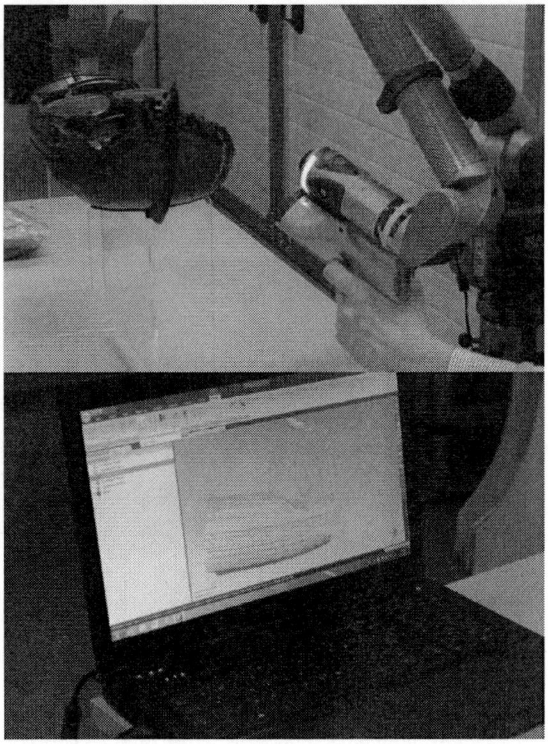

Figure 11: Making a laser scan of the Schuyt model.

6.4 CT SCAN

A CT scan was made of the Schuyt model enabling to see the inner construction of the model (fig. 12).
One of the surprising observations is that the model has a solid stringer mounted on top of the frames just beneath the deck. In still existing boeiers no such construction element is present, although these boeiers have a inner planking just below the deck, supporting the deck beams. This stringer may add to the stiffness and thus compensate for the boeijer's lack of a rubbing strake.
Of course the presence of a stringer out of sight in one model does not automatically mean that this was a common construction, however the drawings that were made in Leningrad before the real boeijer was demolished show exactly the same construction.

Next steps are to combine the CT-scan results with the hand measurements to make a construction plan. Furthermore, it will be explored whether it is possible to translate the CT scan files (Dicom) into a CAD model.

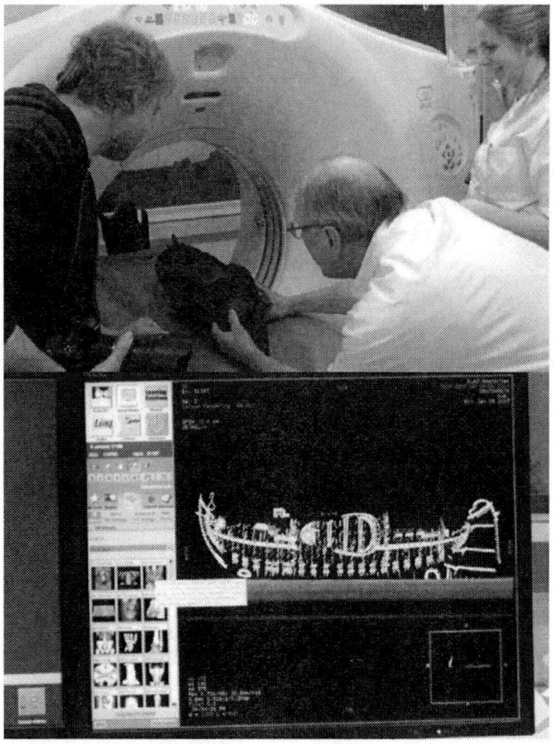

Figure 12: Making a CT scan of the Schuyt model.

7. CONCLUSIONS

The research has shown that the difference between clinker and carvel construction is not the main discriminating characteristic between the 'boeijers' and 'kopjachten' of the 17th until 19th century. The still existing carvel built boeiers are most related to the 'kopjacht'. The boeijers that were clinker built only have overlapping strakes above the waterline, below the waterline the strakes are carvel.

The Boeijer built in 1815 to be presented by the Dutch king tot the Russian emperor was referring to the boeijer bought by Peter the Great in 1697 when he visited the Zaanstreek, the ship building area north of Amsterdam. So the Boeijer can be seen as an early replica.

The story of the construction of the Boeijer for the emperor at the Amsterdam Navy shipyard, as well as the delivery of the Boeijer to the emperor in St. Petersburg is well documented.

Hand measurements and drawing yielded an efficient way to make lines plans of the models of the Boeijer. Comparison of the lines plans of three models indicated that these models were made from the same plans. The plans can be considered to be a reconstruction of the original plans which have disappeared.

Further work needs to be done to process the laser scan data and the CT scan. Furthermore, construction plans may be made as well as hydrostatic and hydrodynamic analyses to serve as a preparation for the actual reconstruction of the emperor's Boeijer.

8. ACKNOWLEDGEMENTS

This project is an initiative of the traditional boatyard "Klaas Hennepoel" and is supported by the Amsterdam Maritime Museum (Mr. P. Rogaar Fellowship 2014 of Het Scheepvaartmuseum) and the heritage programme of the Provincie Zuid-Holland 'Erfgoedlijn Trekvaarten'.

9. REFERENCES

1. SCHELTEMA, JACOBUS, 'Peter de Groote, Keizer van Rusland, in Holland en te Zaamdam in 1697 en 1717' (in Dutch), 1814.

2. WARNSICK, J.C.M., 'De Kweekschool voor de zeevaart en de stuurmanskunst 1785-1935' (in Dutch), 1935.

3. CHEMYSHEV, A.A., 'Navy guard of the Nation' (in Russian), 2013

9. AUTHORS BIOGRAPHY

Alexander de Vos is owner of the traditional boatyard "Klaas Hennepoel". He sailed, maintained and restored traditional Dutch vessels for many years. Furthermore, he researched the history of these vessels. He graduated as a mechanical engineer at Deft University of Technology. After working at TNO as a researcher in the field of traffic behaviour, he switched to restoring the nineteenth century boeier yacht "Vivo". He initiated and constructed the traditional boatyard. In 2014 he was granted the Mr.P. Rogaar Fellowship of the Amsterdam Maritime Museum.

THE SUTTON HOO SAXON SHIP – DEVELOPMENT AND ANALYSIS OF A COMPUTER HULL MODEL PRIOR TO FULL SCALE RECONSTRUCTION

P Handley, Paul Handley Design, UK

SUMMARY

This paper reviews recent research prior to a planned full-size reconstruction of the Saxon ship found at Sutton Hoo. The work included establishing the fundamental form of the ship by reviewing available historic information and data with consideration to possible distortion due to movement of ground and decay of timbers and likely units of measurements used based on measurement of Saxon buildings.

From this research a computer hull model was generated and hydrostatic and stability software used to assess the likely displacement, loading and stability characteristics. Performance under oar power was assessed with performance prediction software and a faired lines plan was generated from the computer model. The Woodbridge River Trust plan to commence construction of the full-size replica in 2017.

1. INTRODUCTION

In 1939 an ancient burial mound was excavated near Sutton Hoo in Suffolk and the remains of an Anglo Saxon ship uncovered. The ship was almost 27m (90ft) long, of wood clinker construction, propelled by up to 40 oarsmen and estimated to date to 625AD.

All of the original timber had decayed, but a well-preserved impression of the plank and frame surfaces was left in the ground and the iron rivets remained in place.

The excavation was cut short by the onset of war and little recorded data survives except for detailed photographs of the ship impression and a provisional lines plan produced by the Science Museum in 1939. In 1967 the site was re-excavated and further lines plans drawn, but by then the hull impression had considerable damage and the excavation team focused mainly on archeological aspects and construction.

The available records and data were reviewed so that an accurate computer model of the ship could be developed to look at the stability and performance characteristics and to provide a lines plan for the planed full-scale reconstruction.

2. HISTORIC DATA

The recorded information, drawings and measurement data available for study for this work included:

- Extracts from notes from the 1939 excavation written by C.W. Phillips and Commander Hutchison, in Volume 1 of the Sutton Hoo Ship Burial by Robert Bruce-Mitford [1].

- Detailed photographs taken in 1939, which provide a clear indication of the overall shape of the ship impression at that time.

- A plan prepared by the Science Museum in 1939 showing the hull lines, section shape at frames and rivet positions for planking. A full copy of the plan is held at the Ipswich Museum and the plan is also shown as Figure 135 in [1].

- Information and observations recorded by Robert Bruce-Mitford and Angela Care Evans following the 1967 re-excavation [1].

- Lines plans of the ship by Colin Mudie, one dated 1973 with hull sections at frame positions [1] and a second dated 1974, number 202.8

- A plan showing planking and rivet positions, including a body plan with hull sections at frames, (Figure 325 in [1]) described as an archaeological plan 'based on 1939 photographic records and information gained by the 1967 re-excavation'.

- Measurement data of the rivet positions taken in the 1967 re-excavation and a permanent record of the 1967 shape taken via a glass fibre moulding taken to the British Museum

No detailed measurement data of the shape of the impression in the ground in 1939 was available, nor any offset tables for the lines plans listed above. Detailed measurement data was taken in the 1967 re-excavation but by this stage the impression had been badly damaged.

3. OBSERVATIONS ON DATA

3.1 1939 PHOTOGRAPHS

From the photos taken the first excavation in 1939, it can be seen that the overall shape was preserved extremely well with the rivets showing the lines of plank edges and some frames showing, although there were some areas of missing detail and local irregularities in shape. Figure 1 is one of several very clear photographs taken in 1939 by

Miss B. Wagstaff. By studying the photos the following points can be observed regarding the shape of the ship:

Extreme ends - It can be seen from the photographs that no detail was uncovered at the extreme ends of the hull in 1939. Figure 2, a photograph of bow taken by C. Phillips, shows this. The loss of the ends was probably due to the proximity to the ground surface and possible ploughing etc. Any drawings showing the bow or stern end extensions are not based on preserved evidence.

Mid-sections - Not much detail of the shape or position of frames can be seen in the middle third of the length of the boat, but considerable detail of frame shapes was preserved forward and aft of this middle section. Therefore, when interpreting the shape shown on any plans less emphasis was given to the middle sections than to the forward and aft thirds.

Bow and stern sections - The photographs in Figure 2 shows that the bow sections of the ship impression were flared out rather than the sharp and more vertical ends normally found on boats of this era, such as the Nydam boat [2]. The same was true at the stern.

The general consensus since the discovery of the Sutton Hoo ship is that this shape at the ends was probably caused by the planks springing away from the stem and stern posts. This could be explained if the sand and soil used to back-fill the cavity outside the hull when it was buried compacted downwards over time, reducing the support of the planks ends near the top of the mound. As the ends were near the ground surface and may have extended beyond the surface, decay of some timbers would have been faster here reducing support and allowing the planks to move.

Topsides - The spreading outwards of the bow and stern planking would have extended at least some way towards the middle of the boat, so throughout the length of the boat there may have been some slight outward movement of the top planks as the soil settled. There seems to be some evidence of this shown in the photos.

3.2 SCIENCE MUSEUM 1939 PLAN

A plan was drawn in 1939 by the Science Museum, part of which is shown in Figure 3. The plan, which was labelled as 'provisional', shows longitudinal lines and sections following fair curves without the local irregularities in shape that can be seen in the photos. It therefore seems reasonable to assume that when the plan was produced, fair curves were drawn as a best fit through the measurement data that was recorded at the time.

Although not stated, it is likely that measurements taken from both sides of the ship impression were averaged for this exercise, thus eliminating concerns over asymmetry, twist or angle of list, which are issues relating to the shape of the ship that were considered by the 1967 excavation team.

The process of creating fair hull lines by hand through a set of unfair points would have been fairly time consuming for the museum staff and they had little time available to do this in 1939 due to the impending war. Accordingly the museum plan's lines are not considered to be a completely accurate record of the ship's shape, particularly given that the ends of the ship are closer to the shape of the impression in the ground than to the shape that might have been expected for a ship of this type and construction.

3.3 1967 RE-EXCATVATION

In 1967 the Sutton Hoo ship mound was re-excavated to establish more information about the ship. Unfortunately the shape of the impression had by then been damaged by military training exercises on the land during the war, but is was still possible for the team to learn more about the ship, in particular relating to construction and areas of detail.

All rivets on the site were collected and preserved after their measurement positions had been recorded and eventually a full-scale glass-fibre casting was made of the impression. This was made in sections to allow transport to be stored at the British Museum. Figure 4 shows the casting assembled.

Although the casting has a boat shape, it is distorted locally and missing much of the top-sides due to the damage to the impression during the war, such that both Rupert Bruce-Mitford and Angela Care Evans stated that the re-excavation did not provide information on the overall shape of the boat that could improve on that recorded by the museum plan.

3.4 PLANS POST 1967

Several more provisional lines plans of the ship were drawn after the 1967 re-excavation, based on the data available from both excavations. At least two lines plans were prepared by the Colin Mudie design office and a further plan was produced illustrating the planking and some other construction detail.

The section lines shown in the Science museum plan were compared with the later lines plans and it could be seen that they all varied in shape. The 1973 Colin Mudie plan showed finer sections throughout the length of the boat than the Science Museum plan. The 1974 Mudie plan showed significantly fuller sections than the 1973 plan, closer to the shape shown in the Science Museum plan. Figure 5 shows a comparison of the 1973 Colin Mudie body plan and the museum plan – the fine lines are from the Colin Mudie plan.

The archaeological plan was somewhere in between the museum plan and 1973 plan, but the hull sections were not faired, so it would seem to be more of an illustration of the likely appearance with some constructional detail.

3.5 HALF-SIZE REPLICA

In 1993 a working half-size replica of the ship, named 'Sae Wyfling' was constructed by Edwin and Joyce Gifford. The lines plans drawn by Colin Mudie were used for the re-construction. The replica was both rowed and sailed, but for sailing the keel timber was extended in depth to improve lateral resistance.

4. DEVELOPMENT OF HULL LINES

4.1 INITIAL LINES BASED ON 1939 PLAN

In order for the 1939 Science Museum plan to be use as the reference shape for generating the initial lines of the ship on the computer, copies of the bow and stern body sections were scanned and printed at larger scale. Some missing detail at frames was added to the printed drawings by fairing curves between points shown clearly on the plan.

This enhanced plan was scanned again and the points representing the positions of the rivets used to join plank edges at each frame station were digitised using Maxsurf Modeller hull design software. The digitizing was done by first importing the scan of the museum body plan into Maxsurf as a background image then inserting 'marker' points over the plan section lines. These digitised marker points were then joined automatically by Maxsurf to re-create the museum body plan as a 2-D image that could be compared on-screen with the lines being generated for the computer model of the hull.

Maxsurf provides a number of options for automatically generating a 3-D hull model to fit a set of hull markers on defined stations. This approach is fine for assessing hydrostatics and stability but as the intention was to produce a fair lines plan the new hull model was generated manually from a small number of control points aiming to ensure fair curves that were a good fit to the marker data.

A computer model of a hull was developed this way and was a reasonably accurate fit to the marker data. If there had been complete confidence with the accuracy of the Science Museum plan then this initial computer hull model could have been presented as the lines plan for the boat. This however was not the case for reasons outlined above, so in order to develop hull lines that better represent the original ship a number of further adjustments were made to the shape.

4.2 ASSUMPTIONS FOR SHIP LINES PLAN

Finer ends - For reasons describe above, an assumption was made that the spoon-shape ends found in the ship impression and shown in the 1939 museum plan do not reflect the correct shape of the original ship. Accordingly the ends of the computer model were pulled in to create a shape that would be expected for planks attached to stem and stern posts as typically found on ships of similar age, construction and style, such as the Nydam boat [2].

Bow section fullness - When the forward sections of the museum plan were compared with the aft sections it could be seen that the forward sections were considerably fuller. There was no measurement data to check, but the difference in fullness could not be seen clearly in the photos.

As a rough check measurements were taken off some frame stations in the photos by measuring the camber of hull segments at these frames. This exercise did not show up much difference in curvature in the bow and stern, so it seems like that the extra fullness in the bow sections was unintentionally exaggerated in the museum plan, probably as a result of expediting the preliminary lines fairing process as discussed above. The bow section fullness was therefore reduced a little in the faired computer hull plan, although the forward sections remain slightly larger than aft sections.

Keel rocker - The computer hull model created to fit the markers from the museum plan had a small amount of keel rocker. This feature was retained throughout the fairing process as it has been suggested by Scandinavian experts that a boat of this type would have had some keel rocker and it is slightly easier to generate a fair hull (or to plank a hull) with a small degree of centreline rocker. This rocker line refers to the line of intersection of the garboard planks on the centreline rather than any additional projecting keel timber.

The lines generated by this fairing process show a smooth and fair overall shape with gently rounded section curves running into a bow and stern shape that may be considered typical for the ship type. Fig 6 shows the hull body section lines at frame stations with the hull markers shown as joined crosses, showing that the shape is quite close to the museum plan at most sections although a fairer set of lines with slightly less curvature in the bow sections.

5. LOAD AND STABILITY ANALYSIS

The faired computer hull model was examined using Maxsurf Stability software to get a feel for the likely displacement, loading and stability characteristics. Maxsurf allows the units to be switched from ft to metres at any time and for this exercise metric units were used even though the hull form was modelled around imperial measurements to reflect the likely original measurement units (see 7.1).

The lines plan, which is shown in Figure 7, represents the shape of ship with the section curves faired through the outside surface of the planks, excluding plank overlaps. This hull shape is a close approximation to the clinker hull shape and suitable for calculating hydrostatics, loading, stability and other performance aspects.

The displacement to the estimated design waterline shown on the Science museum plan, and referred to as the 'probable light load line', was found to be close to 16 tonnes. A rough weight estimate for the ship indicated that this might represent a realistic loading for the boat, as follow:

Area of planking including overlaps is approximately 110 sq. m. For specific gravity of oak of 0.72 and plank thickness of 2.5 cm (1 inch) the weight of oak planking is 1980 kg, say 2 tonnes. If frames, keel and floorboards are taken as 2 tonnes and a further 1 tonne for thwarts, tholes, gunwales, rivets and caulking then a rough estimate for the weight light-ship without ballast is 5 tons.

If the crew number is taken at 40 with average weight of 100kg each including oars and some kit, then the crew adds another 4 tons. If 10 passengers are carried this might add another ton, bringing the light load displacement to 10 tonnes.

If stores/cargo of 3 tonnes were carried and possibly 3 tonnes of ballast, then the total loaded displacement would be around 16 tonnes displacement which coincides with the estimated waterline on the Science museum plan. At this load the canoe body depth at mid section is approximately 2 ft and midship freeboard also 2ft, which might suggest that this is at the upper end of the load range rather than a 'light load' as stated on the museum plan.

The hydrostatics were run at a series of waterlines and the load range from 10 to 16 tons looked appropriate for the design. As an absolute limit, the displacement when loaded to the point that the gunwales were at the waterline was found to be 50 tonnes. Given the very light construction of ship it seems unlikely that it would have been used for heavy load carrying and this also raised the question as to whether or not any ballast would have been carried if it was solely powered by oars, as light weight would have been beneficial for performance.

Stability was checked for the range of load cases from 10 to 16 tonnes based on estimated vertical COG for the loads listed above, with no cargo load or ballast included for the 10 tonne load case. At each of these load cases the stability was substantial, such that ten persons, for example, could all move to one side of the boat at the same time with only a small angle of heel resulting.

A large angle stability assessment indicated that the hull could heel to 20 degrees at a 16 tonnes load before down-flooding, suggesting an ample reserve of stability for rowing but at the same time that it was unlikely that the boat was intended to be sailed at an angle of heel. So if any sail was used it would most likely have been for downwind sailing only. In general though, the characteristics of the form seem to be more in line with that of a large rowing vessel rather than a sailing boat.

6. HULL FORM AND PERFORMANCE

6.1 HULL FORM

The hull model developed is fairly symmetrical fore and aft although the forward sections are slightly wider and fuller than the stern sections, such that the longitudinal centre of buoyancy to the reference waterline is 2% of the waterline length forward of amidships. A longitudinal centre of buoyancy (LCB) forward of amidships is not unusual for historic ships, and given the potential advantage of greater buoyancy forward keeping the bows up in rough weather or for beaching in surf, it seems appropriate for the design.

The rounded hull sections are perhaps to be expected if the boat was designed for rowing as round sections reduce wetted surface area and hence drag, as well as achieving relatively low draught for any given load which would be helpful for operating in shallow water, such as restricted tidal areas.

The hull has a small amount of fore and aft rocker on the centreline which again is perhaps to be expected for a rowing craft, as it both reduces the wetted surface area in the ends and improves manoeuvrability in restricted areas. Some rocker would enable the bows of the boat to get closer to the shore edge when running up a beach to enable crew to disembark without getting too wet. Further, a small amount of keel rocker would enable the boat to be freed and turned more easily after grounding. So unless there was a need for deeper ends and keel depth - for lateral resistance for sailing - it seems reasonable that these features would be absent and the hull would have a small amount of rocker with shallow keel timber.

6.2 POWERING AND RESISTANCE

The prismatic coefficient for the model was approximately 0.6. This is towards the high end suggesting a hull capable of reaching a displacement speed of around 9 to 10 knots for a hull of this length and relatively low power from rowing.

The performance of the hull with respect to resistance, power and likely speed was further assessed for the final lines using both empirical and calculation methods in Maxsurf Resistance software. Power estimates for a range of speeds were calculated at 13 tonnes displacement by three empirical methods (Holtrop, Fung and upright resistance from the Delft III sailing VPP) and by applying slender body theory, which is applicable for the length to beam ratio and displacement of the ship. It can be seen from Figure 8, which shows the predicted power curves, that the methods compare quite closely over the expected range of speeds. The approximate predicted power is 10hp at 8 knots, 20hp at 9 knots and 30hp at 10 knots.

It is difficult to estimate the power delivered by a human with an oar, and also there is a question over the number of oarsmen, which may have been 28 or 40, depending on whether or not tholes for oars were originally fitted in the mid section of the ship but removed for the burial.

6.3 SEA-KEEPING AND STRUCTURAL INTEGRITY

As a further exercise the computer hull model was animated in waves of 1m height and 2m height. In waves of 1m it could be seen that the freeboard remained almost uniform as the bow and stern lifted the hull to the waveform. At 2m the ship appeared to be taking some water aboard at the ends, which could have been deflected by temporary decking. However, it seems unlikely that the ship would have operated in large waves as it's lightweight construction and low amidships freeboard may well have lead to structural issues.

As a further exercise the structural strength of the ship could be assessed to establish what the maximum load and wave height combinations would be whist still maintaining a good margin for structural integrity.

7. CONSTRUCTION

7.1 SCANTLINGS AND MEASUREMENTS

Measurements of the frame positions taken from the plans showed that the average frame spacing, ignoring irregular spacing at the ends of the boat, was very close to 3 feet (2.994 ft average).

This 3-foot frame spacing may be coincidental, but a review of measurements used in Saxon times for land buildings indicated that it was likely that there were two measurement systems in use at the time.

One was based on a 'rod', which was 15 German feet long with each German foot 13.2 inches long, so the German rod was 16.5 modern feet long. The other was based on Roman units where a 'rod' was 15 feet long with each foot 12 inches long, these feet and inches being the same size as modern imperial units.

It therefore seems a reasonable assumption that the Roman/current imperial measurement system was used for the Sutton Hoo ship, particularly when further noting that the overall distance between the end-most frames is 75 feet or 5 'rods'.

Also, that the average planking thickness is 1 inch based on measurements taken from plank scarf join rivets. And while it could be coincidental, it was noted that the average width of the hull planks in the mid sections is approximately 1 ft - see 7.2 below.

If this measurement system assumption is correct, then it would suggest that there has been no significant expansion or contraction of the ground over the 75 ft length between the end frames, and that the overall dimensions of the impression may be considered very accurate, perhaps more so than the dimensions of preserved historic wooden ships where timber distortion and shrinkage can lead to considerable speculation over the original ship dimensions and shape, as is the case for the Nydam ship, as shown in Figure 10.

7.2 PLANKING AND FRAMING

The plank runs as recorded by the rivet positions shown on the Science Museum plan were represented on the computer by marker positions. There were some discrepancies found in the shape of the longitudinal lines passing through these points, which represent the plank overlap paths, particularly in the mid portion of the boat, which is perhaps to be expected due to the lack of detail uncovered in the mid section.

The plank widths (from plank edge rivet to rivet) were recorded at longitudinal positions during the re-excavation and although this information is also mostly missing for the mid portion of the boat it does give a good idea of the average plank width, which is around 1 ft, excluding plank taper in the boat ends. This is based on 10.5 inch average distance between the edge rivets plus 1.5 inch plank overlap.

As a further check on plank width the girth of the midship frame section of the computer hull model was measured and it was found that the correct girth would be achieved by 9 clinker planks each of 1ft width with overlaps around 1.3 inches. This was based on a half-girth measurement of 8.5 ft less a small amount for the keel timber width and a slightly wider sheer plank.

The plank timber is thought to be oak. The planking would have originally been made by splitting trees and trimming the planks with adze or axe [3]. The intention is to attempt to produce planks in the same way for the reconstruction of the ship.

The lines plan derived from the computer model shows a frame spacing of 3ft except for the end frames beyond the drawn waterline, so that the frame positions will coincide closely with the original frame positions. A nominal design waterline was drawn with a draught of 2ft which gives a waterline length of 60ft with the forward perpendicular (FP) at Frame 3.

8. CONCLUSIONS

Despite the complete absence of any of the original timbers, it can be concluded that the impression in the ground provided a remarkably good record of the ship's shape, perhaps even more reliable than historic ships that are found intact but subject to indeterminate distortion and shrinkage of the timbers. Such was the preservation of the layer between the timbers and surrounding sand and soil, and the 'map' of rivets, that it was also possible to establish considerable construction detail too.

Taking into consideration the available recorded data on the shape of the impression and subsequent analysis, a fair computer model of the hull has been generated. In addition to providing a lines plan for the planned reconstruction of the ship this model provided useful information on stability and likely performance.

The Woodbridge Trust plans to start construction of the full-scale replica in 2007 at a special site in Woodbridge, where the boat will eventually be based for public viewing.

9. ACKNOWLEDGEMENTS

The Author would like to thank the Woodbridge Riverside Trust and in particular Paul Constantine for proposing the project to create a computer model of the Sutton Hoo ship and for providing regular input to the project. Thanks are also due to Peter Bradbeer and Joe Startin for their enthusiastic input.

10. REFERENCES

1. BRUCE-MITFORD, R., Volume 1 of The Sutton Hoo Ship Burial, British Museum, 1975

2. ABEGG-WIGG, A., 'Das Nydamboot', 2014

3. CONSTANTINE, P., 'Anglo-Saxon Craft Construction', 2015

11. AUTHORS BIOGRAPHY

Paul Handley MRINA is an independent consultant Naval Architect and small boat designer. He was a consultant to the European Standardisation Committee for eleven years assisting with development of ISO standards supporting the Recreational Craft Directive and continues to provide consultancy on boat safety standards and design. His boat designs include two ISAF international youth sailing classes, the RS Feva and RS Tera, several adult dinghy and keelboat classes and rowing sculls.

Figure 1: The ship trench in 1939, photograph by Miss B. Wagstaff

Figure 2: Bow sections of the ship impression, photograph by C. Philips

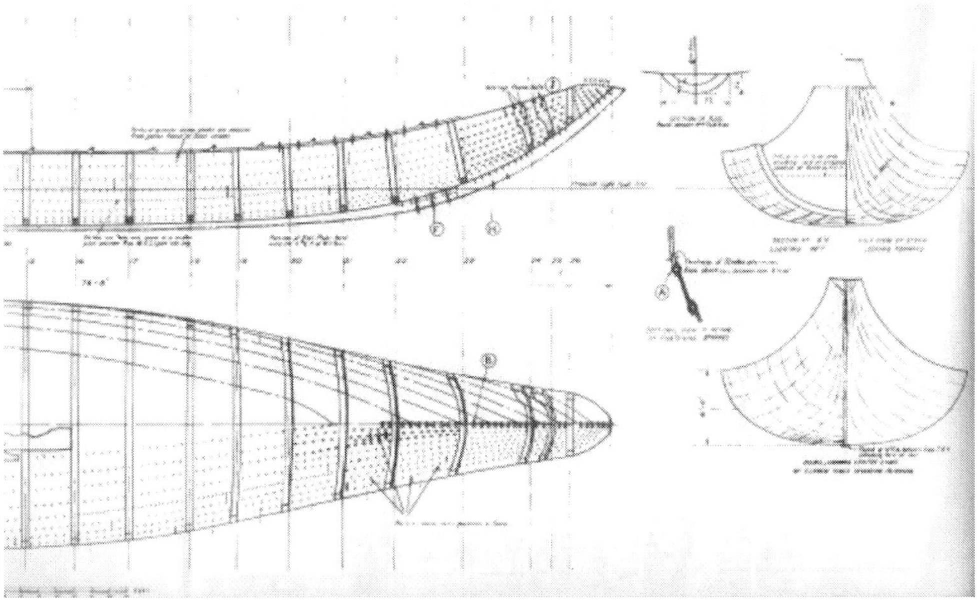

Figure 3: 1939 Science Museum plan

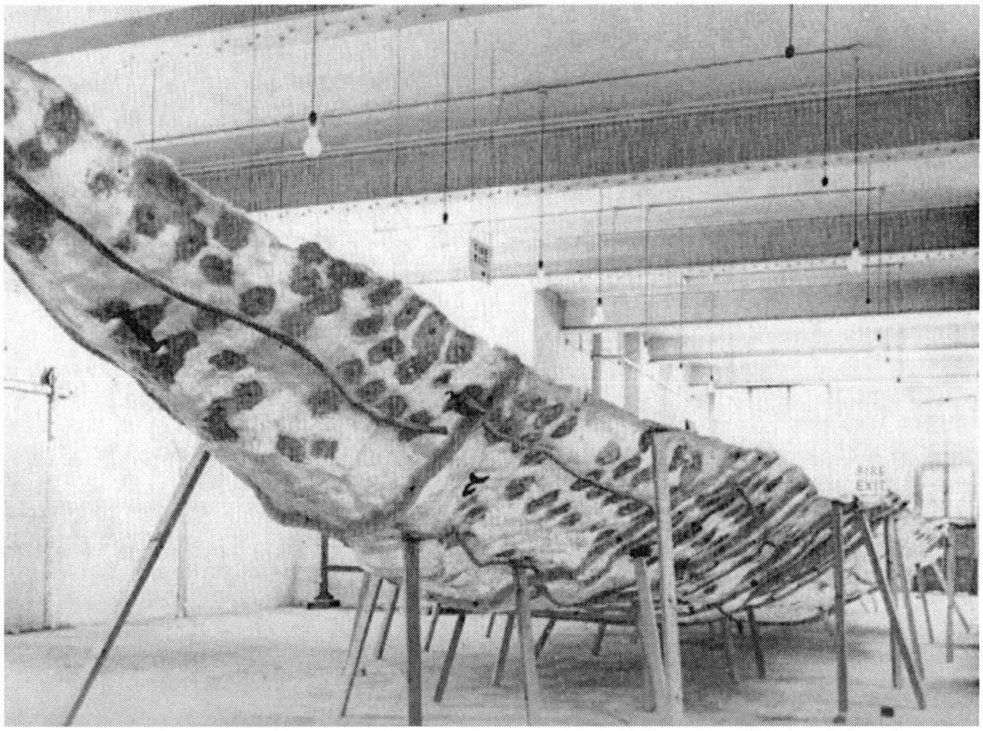

Figure 4: Casting of the 1967 ship trench, photograph by BBC

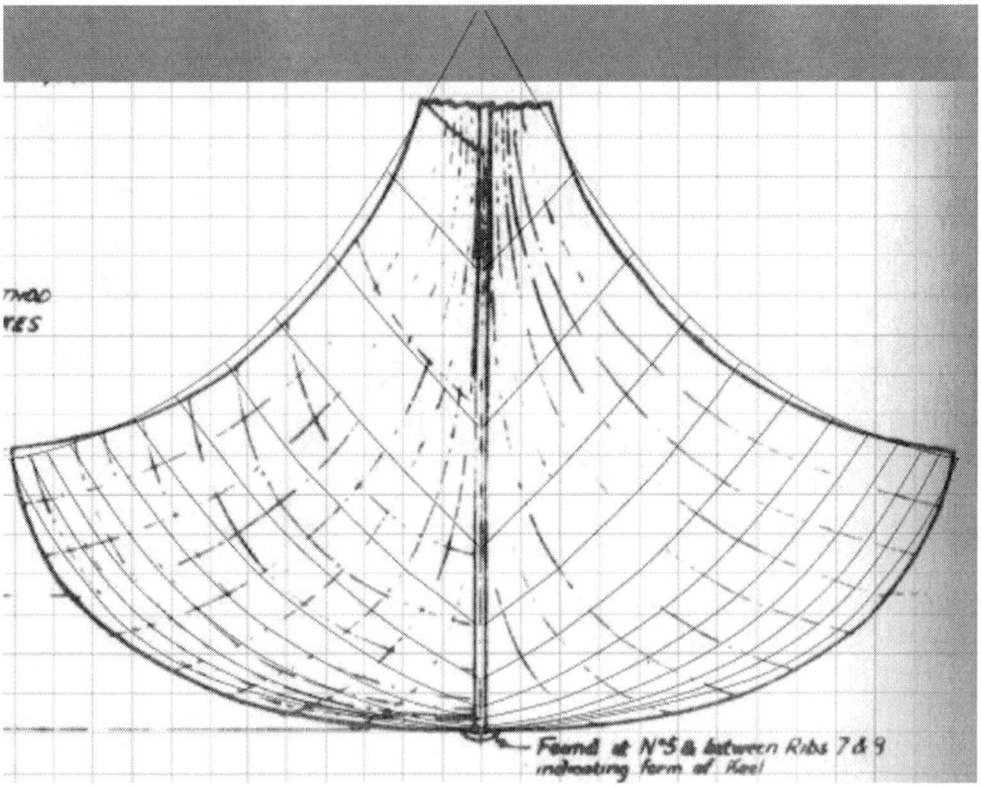

Figure 5: Comparison of body plan by Science Museum and Colin Mudie (1973)

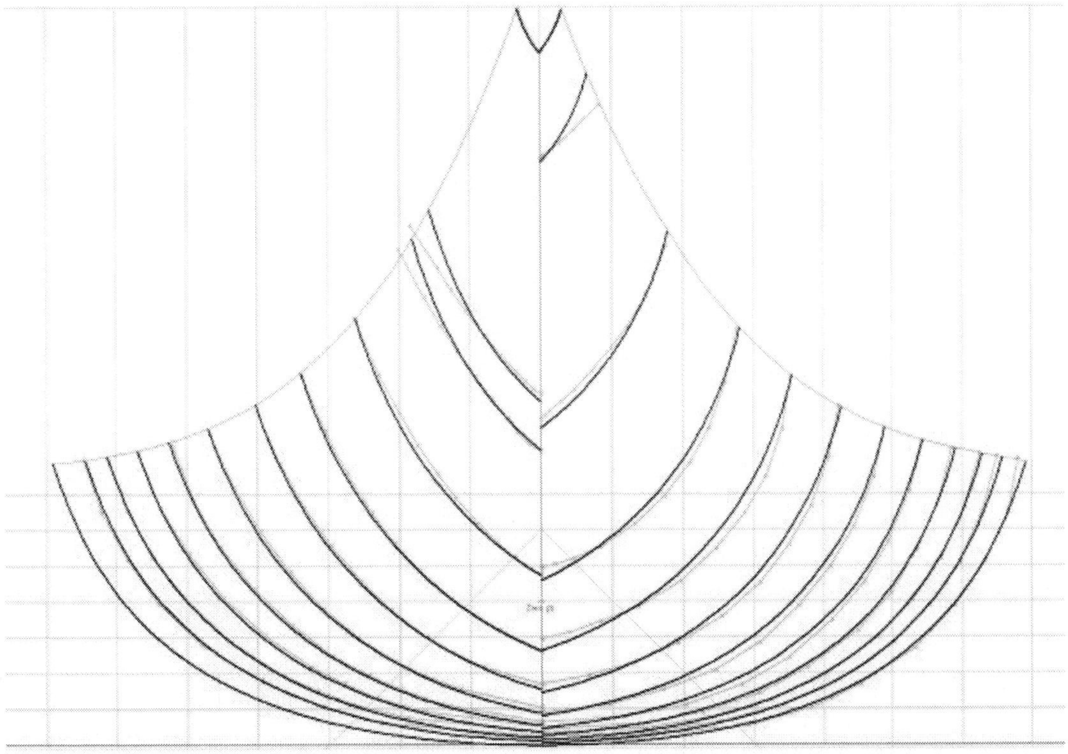

Figure 6: Computer body plan lines compared to Science Museum lines (through makers)

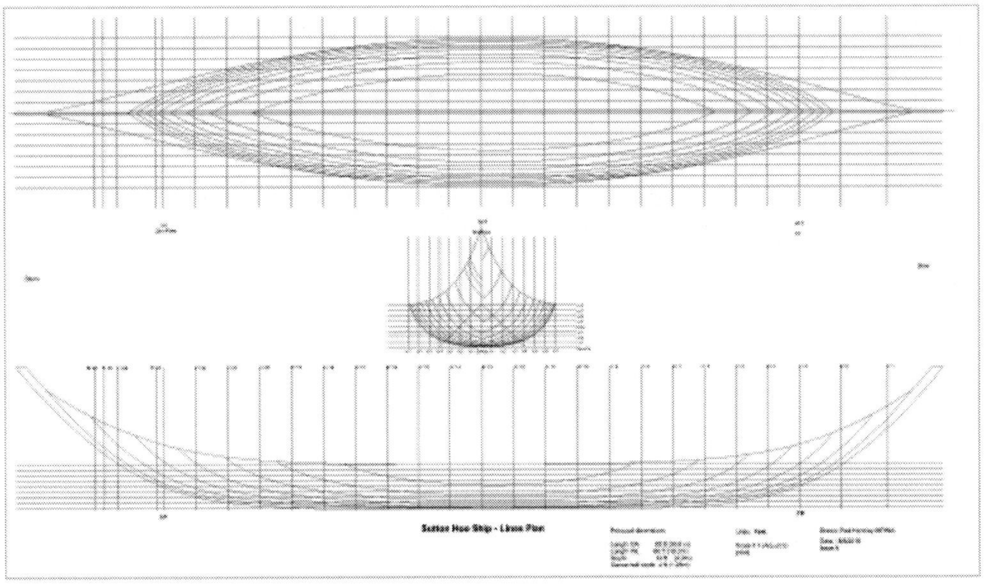

Figure 7: Lines plan from computer model

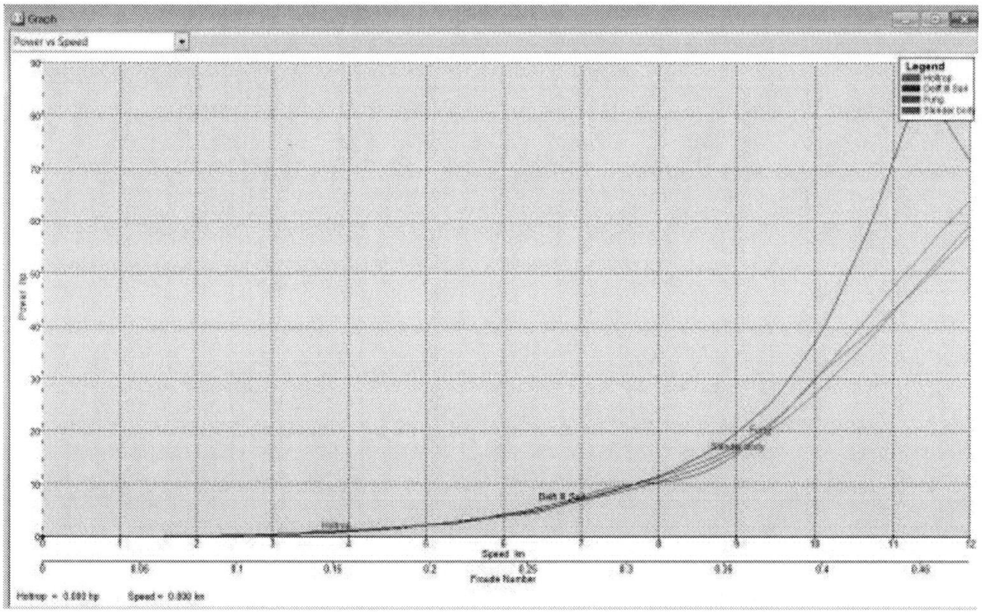

Figure 8: Predicted power curves for Saxon ship

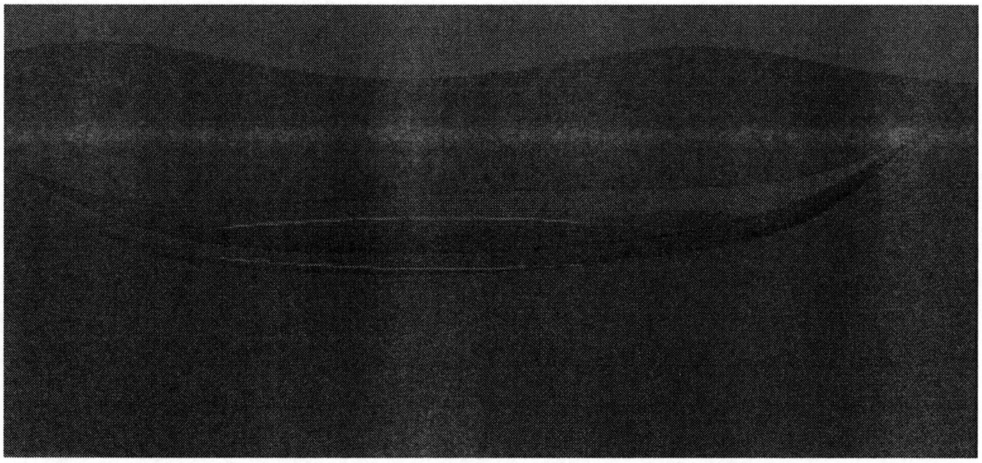

Figure 9: Ship in 1m waves

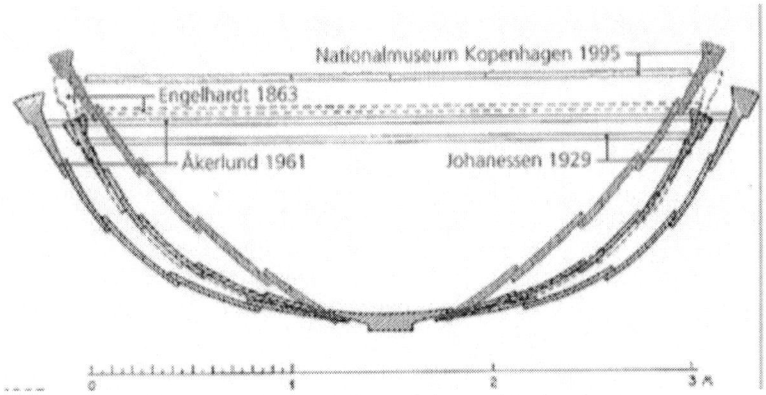

Figure 10: Interpretations of shape of Nydam boat

LILIAN: FROM GENTLEMAN'S YACHT TO SCRAP AND BACK

R G White Norway and **R S Pereira** UK

SUMMARY

The gentleman's motor yacht *Lilian* is celebrating its centenary this year. Built in Sweden for a Danish banker, *Lilian* was the largest yacht designed by the Swedish architect C G Pettersson. It was conceived as an ultra-modern vessel and designed around a handed pair of lightweight Polar Z-type diesel engines.This paperreviews the design features of this pioneering steel diesel-powered yacht with reference to surviving Pettersson detailed drawings, and covers the challenges of maintenance and the choices between originality and operational reality and restoration or reconstruction.

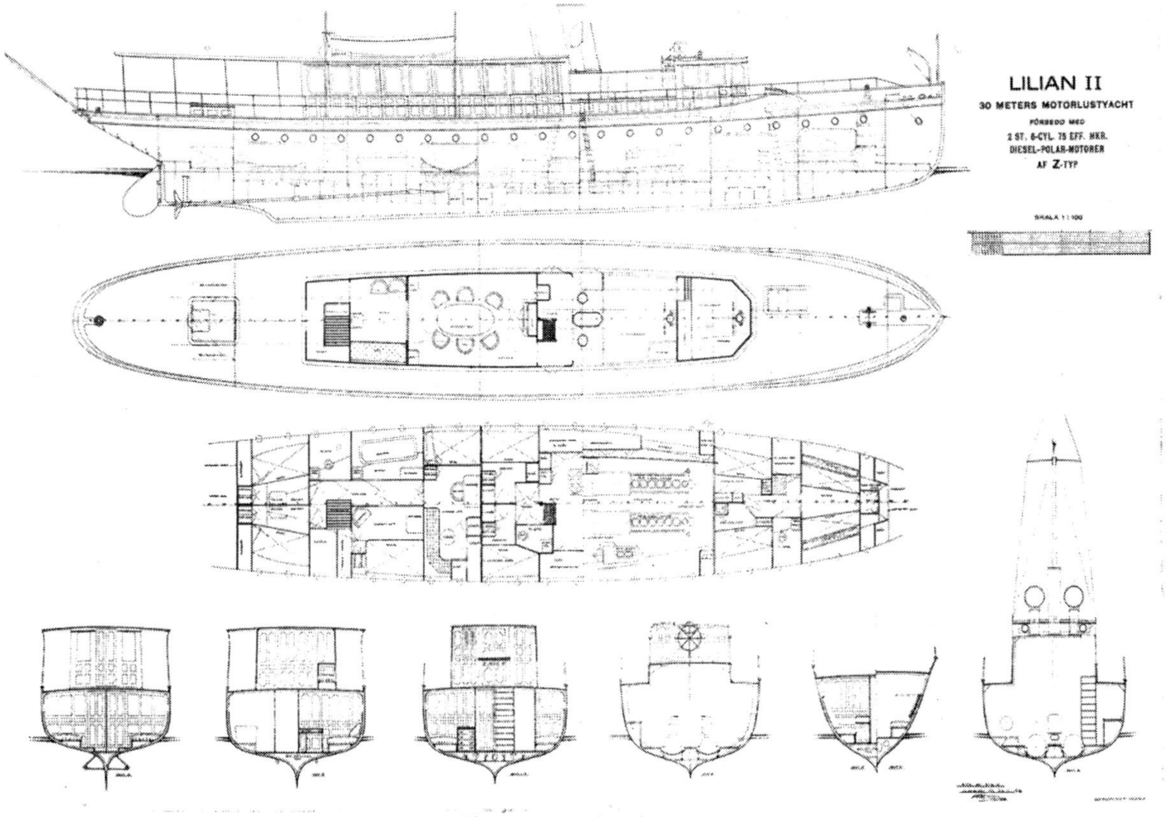

Figure 1: GA drawing

1. INTRODUCTION

When the motor yacht*Lilian* was handed over to its first owner it is unlikely that the owner, the yard, the engine builder or the designer would have imagined that a hundred years later the vessel would not only be providing a permanent home but also making periodic voyages in reasonably settled weather in the range from the Baltic to the mouth of the English Channel. That first ownership ended in scandal but for many years *Lilian* fulfilled the functions of a pleasure yacht, then slowly declined into unseaworthy condition.

Built for the Danish banker Emil Glückstadt, after the collapse of his bank and his death *Lilian* was owned from 1923-1926 by Mr Huntly-Gordon from Poole. From 1926-1935 it was owned by René Bolloré, renamed *Dahu II* and registered in Quimper, France. The next owner, Mr Ferguson Roydhouse, moored the yacht on the Thames in Kingston 1935-1967.Thereafter it became the static HQ of sea cadets, renamed *TS Windsor Castle*. The present owners acquired *Lilian* in 1980.

For the design Glückstadt went to the Swedish naval architect CG Pettersson, though *Lilian* is not typical of Pettersson's output. CG Pettersson has achieved almost iconic status as a designer and the phrase 'Petterssonboat' conjures up a picture of an elegant and easily driven motor boat, often in varnished mahogany [1]. Many of these survive, valuable enough to warrant expensive restorations. He designed a number of larger commercial vessels, for example the 'beer boats' used by a brewery to distribute beer and soft drinks to restaurants around the Stockholm archipelago. But it seems that *Lilian* is the largest Pettersson vessel to survive.

The yacht was built in Stockholm at SödraVarvet, formerly a large and influential shipbuilder. The site of the yard is now the Viking Line ferry terminal. *Lilian's* hull form took advantage of the comparative compactness and light weight of the diesel engines, which were a pair of Polar type Z units made by AB Diesels Motorer at Sickla in the same south east part of Stockholm as the shipyard.

Quite lightly built in riveted steel, *Lilian* has angle-steel frames at 0.5m spacing with deeper web floors and five watertight bulkheads. The teak deck runs in a smooth sheer and camber from bow to stern but the visual sheer of the deck edge is broken, a visual device very typical of Pettersson designs. Pettersson was able to use a shallow canoe body hull form with a wine glass section anda comparatively deep keel. Overall dimensions are 30m long, 5m beam and draught aft about 2m. The engines are also mounted quite far forward leading to exceptionally long propeller shafts with half the length outside the hull supported by A-brackets.

The overall layout can be seen from the GA drawing (see Figure 1). At the extreme bow is the chain locker and collision bulkhead. Aft of this comes the former crew accommodation with the wheelhouse above. The engine room has a raised poop deck over it and a further raised section carrying the funnel and housing the fuel day tank, the water tank and the engine room access ladder. The main accommodation is in the hull and the panelled wooden deckhouse. Aft of this under the stern deck is the lazarettehousing four fuel tanks, two large and two small, the rudder trunk and steering gear and storage space. Access is by a companionway on the aft deck.

2. FIRST ENCOUNTER

When the current owners first came across *Lilian*, it had just hit Richmond Bridge on the way from Windsor 'to the South of France to be chartered'. It was tied up on Corporation Island like a great beached white whale, with an invitation from the owner to come aboard and see how the engines worked, in case there was interest in buying her. Never reluctant for a challenge, he was watched going through a complex ritual involving diving bottles and flexible hoses off building sites, the ancient engines finally bursting into life.

In spite of the evidence of massive neglect, the mess, the mustiness and the rickety decks, there was a nice old lady underneath – obviously needing some cosmetic work and a face-lift but the elegant bone structure was still apparent. The current owners were actually looking for such a boat and the dimensions were almost exactly those of their riverside mooring at 30metres long and draught under two metres. But they shook their heads and made excuses and left.

A year later, *Lilian* was sitting on a mooring just down-river, propped up against some concrete pilings at the Gasworks wharf in Brentford (figure 2). So a friendly surveyor was found and *Lilian* was bought, for one pound plus other considerations, and then towed to the home mooring in Twickenham.

Figure 2: *Lilian* in 1979

The sea cadets who had used *Lilian* as a training ship at Windsor had left their mark (see Figure 3). The varnished mahogany superstructure had been painted white,and much of the accessible lead ballast, the telephones, the binnacle and even sections of the copper central heating pipes had gone. Where they needed to hang ropes, six-inch nails had been hammered into the panelling, and initials and worse had been scratched liberally throughout, with one sealed bulkhead crudely cut open as a serving hatch. Every porthole glass had been smashed. The dynamo and generator engine had been immersed too often to work reliably, and the 110volt set of batteries had vanished. So the owners and two young children lived for a while with a festoon of building site light bulbs, a garden hose for water and sleeping quarters in the few bunks that were not under drips from the leaking decks.

Figure 3: Sea Cadets (late 1960s)

3. REPAIR AND RESTORATION

The forecastle was the worst. The teak planks were laid on steel beams that were so corroded that not one spanned the deck intact. The interior here had mostly been gutted, and what was left was rotten timber of poor quality knocked together originally for the crew's quarters. So there was little compunction in removing what was left, lifting the teak planking (carefully numbered for re-use) and laying a steel deck, with access to the forecastle made via the wheelhouse. Steel decks were extended around the wheelhouse and down both side-decks, including a new bridge and steel wheelhouse base (see Figure 4). At a stroke, *Lilian* was both strengthened and watertight from above, and once the saloon roof had been stripped of several layers of rotting canvas, even that section became dry.

Figure 4: Replacing port side deck with steel (1981)

The mahogany gunwale capping, bolted down with inserts for each stanchion, was almost all replaced, using Brazilian mahogany, including some large curved sections at bow and stern that were prodigal with timber, and some that took weeks to steam and cramp into place.To avoid potential water damage from surface plugs in bolt holes, it was opted to use galvanised coach screws from below, which has proved robust enough in practice.

The inch thick teak decking had been laid directly onto the steel beams, fastened by steel screws from below, with no visible caulking seam but a groove in the sides of each plank (below the surface) that had a string of caulking cotton. After several years with steel decks, the same visual effect of the original was restored by laying marine ply onto the steel using levelling Sikaflex mastic, and gluing the teak strips with a small number of plugs. While the convention for motor boats is to lay parallel straight deck planks, *Lilian* had the much more complex swept teak planking, tapered at bow and stern, characteristic of sailing yachts. The large slabs of timber under the winches at bow and stern were also replaced.

The surveyor, Frank Bandey, who had condemned *Lilian* as unseaworthy twenty years before, said that with a new bottom he could see it fit for the Channel, even the North Sea. So with perhaps unwarranted optimism *Lilian* was consigned to a local boatyard slipway (Strong Marine on Eel Pie Island, Twickenham) to have a new bottom.

They started with a solid bar keel, then shaped plates from there to the waterline, welding through to the old frames where they could. Lilian had started with riveted steel plates, probably only 5mm or less (<u>much</u> less by the time she was over-plated) on rather light frames. A photo from the original builder's yard shows an elegant structure with rolled plates being bolted, then riveted in place (see Figure 5). The long engines with their gearboxes were bolted to full length bearers in the spacious engineroom.

Knowing that condensation and insulation was likely to be a future issue, it was elected to shot-blast and prime all exposed steel work down to the water-line and spray with polyurethane foam, then hide it behind wooden panelling, either the replaced original or replica to match. As part of the dismantling, with regret the mass of wiring that supplied lights, bells and telephones was replaced – most of this was thin tinned copper wrapped in fabric inside lead sheaths.As it was thought 110v DC systems were going to be too hard to replace, cabling was changed to 24v throughout (apart from the engine room motors, which went to 240v AC). The fine electrical charging board, mounted on two slabs of marble, was removed and consigned to store and the old dynamo (110v 4.5kVA) was considered beyond repair. Even an old Garrard 78rpm electric gramophone fitted in the saloon had to go.

Figure 5: Construction 1916

On tackling the broken port-hole glasses, it was discovered that all the (42) pairs had stamped identifiers – and many were in the wrong places. When matched up, they fitted more snugly, and most became watertight. The bronze frames were riveted into the steel hull-sides with countersunk steel rivets(sourced in Birmingham), the seals were rubber (sourced from an ancient stock at Norris of Isleworth) and the wooden surrounds were replaced with exact fibreglass replicas since all the old wooden ones were rotting. A puzzling set of a dozen galvanised steel objects turned out to be scuttle ventilators – designed to be inserted into port-hole apertures to increase air-flow through the accommodation in hot weather.

There had been central heating from new, and the piping diagrams evidence that they used the same pipes to circulate hot water to the taps. Unfortunately, what remained of the heavy-duty Imperial-sized copper pipes and fittings were impracticable to mate to modern metric systems, so the lot was re-piped.

There were three sea lavatories on board, but because of the potential pollution to the river, the plumbing was arranged such that effluent could be switched to the original ballast tanks, and for these to be pumped out intermittently to the mains sewer on land. The tanks were first cleaned out (via the large bolted manhole covers) and internally sprayed with a rubberised coating. The sea connections had been sealed by the new hull plating, and were not re-used. As the tanks had a considerable capacity, it made sense to drain all grey water (baths, showers, washing machine, dishwasher, basins) into these as well, which had the advantage of using gravity drainage.

Once the services were in (heating, lighting, plumbing), the restoration of the interior was contemplated. The mahogany panelling used almost all the same moulding pattern throughout, and this was fitted precisely and pressed in, rather than pinned. Although this seemed unnecessarily elaborate (particularly as there were almost no square or right-angled corners), it had the advantage that when the superstructure moved it was flexible enough to avoid cracking or breaking the many fitted mirrors.Much had been painted, but once the paint was stripped, the French polishing beneath could be revived and restored. This finish has been used throughout rather than varnishing, as it seems to survive and revive better for interior woodwork.

At each stage came decisions between maintaining originality and practicality in a world which has changed so dramatically since 1916. To take a few examples, the main accommodation is very original, whereas the forecastle accommodation was completely rebuilt as a mini flat for the owners' daughter combined with crew accommodation when cruising. The space is now normally used for guests, or crew when on a voyage. The main galley, once characterised by the widow of a former owner as being squalid in her day, has been rebuilt several times with new equipment, and is cramped but very functional.

Original furniture in the deck saloon includes the large table and the chairs and sideboard. At its aft end the saloon opens through sliding doors into a music room with a piano, labelled as the smoking room on the original plans. Access to the accommodation in the hull is by steep steps from the deckhouse lobbies forward and aft.The furnishings were mostly long gone, but there was evidence of the screwed fittings for carpeting the accommodation throughout, with some old linoleum-like coverings in the galley; the removable screws for the saloon table were found and restored after the first time it fell over at sea. The dumb waiter from the galley to the saloon had lost its massive lead slab counterweight, but it was found in the bilges and re-attached to slide in its close-fitting housing via the large pulley re-attached to the deck-head (Figure 6).

Figure 6: The dumb waiter

4. MACHINERY

Lilian's machinery is a saga in itself [2].For the past few years the original Polar Z engines have not been installed, for good reason. But they are preserved and can go straight back into the yacht when conditions are appropriate.

4.1 THE ENGINES

The Polar twin diesel engines are 6-cylinder units each developing 75hp at 550rpm. An offshoot of Atlas, AB Diesels Motorer(ABDM) is believed to have been the first factory established solely to build diesel engines, as opposed to the many existing engineering works which took licences to add diesel engines to their product ranges, and from 1909 sold its engines under the Polar tradename. ABCM's chief engineer/designer was Jonas Hesselman, who had joined ABDM in 1899 as a young engineer and quickly progressed. Two technical developments in particular made his name in diesel engineering technology worldwide: - the bowl in piston combustion chamber and the so-called Swedish nozzle for blast injection fuel systems. He was a strong believer in higher engine speeds and numerous small cylinders.

This then was the background to the introduction of the Z type engine in 1912, a range intended to cover the widest variety of applications with a compact unit for land power generation, marine auxiliary, propulsion of smaller vessels, industrial power and railcar propulsion. Engines with one to six cylinders were offered with a uniform rating of 12.5hp per cylinder and irrespective of the number of cylinders the engine incorporated a two-stage compressor for supplying air for blast injection and for starting. They were delivered as bare engines, with attached generators, or as compressor sets for air supply.

Six cylinder engines such as those in *Lilian* were popular from the start and a pioneering application was in the first successful diesel electric railcars, built for the Swedish secondary network of railways and covering their traffic requirements more economically than steam locomotives.

Lilian's two 6 cylinder engines are handed, both physically and in rotation, with the camshafts facing the central alleyway and the exhausts and oil coolers on the outside of the engines (figure 7). Individual cylinders are mounted on a common crankcase housing the camshaft, which in turn sits on a common bedplate carrying the crankshaft. Pushrods and rockers operate the overhead valves, the airstart valves on the first three cylinders, and originally the fuel valves. There is no inlet manifold, the air being drawn though the valve cages. Individual watercooled exhaust elbows are linked to convey exhaust gases to the funnel.

Figure 7: Starboard Polar Atlas engine

4.2 THE GEARBOXES

Various options for providing marine diesel engines with reverse were in use in the early days – direct start, CP propellers, electric or electro/mechanical. *Lilian* is equipped with mechanical gearboxes using clutches and an epicyclic train to obtain astern propeller rotation. This was not a new idea in 1916, but it was not easy to design a good one. The gearbox is a mighty piece of equipment, almost the same length as the engine itself. It is divided horizontally on the crankshaft centreline, its lower half is a semi-circular cast iron trough with a mounting flange while the upper half is divided into sections. The upper half forward end is a short U-shaped casting, at the aft end is a much larger casting of the same profile which carries the operating mechanism.
Unusually, the thrust block is between the gear cluster and the engine. Its bearing is carried in a substantial cross-member casting which bolts to lugs cast in the lower gearbox half. At the aft end of the gearbox the output shaft is carried in a white metal bearing and terminates in a taper for the propeller shaft coupling. At the engine end the input shaft terminates in a flange and the heavy flywheel is sandwiched between the engine crankshaft flange and the gearbox flange with bolts running through all three components.

A large drum houses the reverse gear itself. Ahead rotation is obtained by clutching the input and output shafts together giving direct drive, the clutch having alternate bronze and steel plates. For neutral the clutch is freed and to go astern a band brake strap is tightened to lock the rotating drum and bring the spur gear epicyclic train into action to reverse the direction of the output shaft.

To actually obtain these operations involves a long train of mechanical components on *Lilian*, beginning with two impressive bronze levers sprouting from the deck alongside the steering position on the poop deck above the engine room. Two levers side by side work in a notched quadrant, each controlling a gearbox. Rods and chains carry the motion to a chain wheel on the side of the gearbox.

Like many mechanical gearboxes since then it works well when perfectly adjusted. But there is a fine margin between the ahead clutch slipping and heating up, so warping the plates, and a tighter adjustment which requires superhuman strength at the deck lever to force the gearbox into engagement.

These gearboxes show signs of many of the details being worked out actually on the job and after the main casting was made. In particular the oil pump which serves both engine and gearbox is driven by an eccentric mounted on the shaft just aft of the thrust block. The pump body itself is bracketed to the gearbox lower half casting. However, both the oil pump mounting and several of the lugs for band brake and thrust block attachment are wrongly placed and drilled off centre and the pump mounting shows many signs of filing to get the components to fit.

It is perhaps unfair to criticise early mechanical gearboxes which have given service over thousands of miles and a hundred years of intermittent use. However, it seems that ABDM was having second thoughts about the design because among the drawings are some basic outlines for an entirely different mechanical gearbox, a much more compact unit. Unfortunately, the unit is not shown in detail but it is evidently also an epicyclic box, with the thrust bearing mounted in the aft end of the box on the output shaft.

Lilian's steering gear was a chain and cable system running through the accommodation (through several sealed bulkheads) from the bridge to a large quadrant under the stern deck. The shaft and gearing for the forward steering position in the wheelhouse had gone (as had that wheel). Although it was in good working order, advice was that change should be made to a Wagner hydraulic system, both to make the dual steering positions easier to re-fit, and to make autopilot connection feasible. Although there is a chart table large enough for full-size Admiralty charts, there are now electronic charting systems as well.

5. RESTORATION OR RECONSTRUCTION?

In the early years, imagination and what was still left was used to restore the structures. The sea cadets had done a plan drawing of the interior as they found it. However, while in Stockholm for a scientific conference, Hilary visited the maritime museum and enquired of a curator whether there were records of the boat. By chance, that curator was Åsa Fritz-Crone, the wife of the maritime historian Gunnar Fritz-Crone, who was writing a book about CG Pettersson [1]. The SödraVarvet shipyard records were being recorded on microfiche and the original blueprints of *Lilian* were available. So Hilary returned with copies of over fifty original drawings (including Pettersson's draft designs) and suddenly there were templates to use for research and restoration. Years later, a cache of Pettersson drawings, rather surprisingly including wood panelling diagrams, were found in the records of the engine manufacturers (the Atlas engine branch was by then part of Wärtsilä), stored in archives in Gothenburg. These included details of the two ship's boats, the boarding ladder and even the elegant design for the transfer lever in the main head that switched the flushing water inlet from the washbasin to the lavatory.
It was then realised that the two masts, based on a Pettersson drawing acquired in the first batch of records, were incorrect. A photo emerged which showed at launch a single mast behind the funnel (Figure 8).

Figure 8: The mast arrangement, as on one design and as reconstructed (above), and at launch (below)

The multiple draft drawings had shown numerous mast and accommodation arrangements, presumably to be selected from by the owner. Dates on the drawings indicated the finalised versions as used by the boatyard. The two masts had been built, rigged and used successfully for some years, and they were a much more practical rig, so have been kept, in spite of some historical inaccuracy. The schooner rig also looks and feels well balanced.

When on the topic of non-original upgrades, several types of stabiliser have been considered but other restoration work and upgradings were prioritised. Normally *Lilian* has a comfortable motion but can roll heavily under certain sea conditions, which is not only uncomfortable but puts a severe racking stress on the wooden deckhouse. Active fin and passive tank stabilisers were investigated but rejected for various reasons. Gyroscopic stabilisers were a possibility and a thorough investigation was made of bilge keels which would provide a roll damping effect both in motion and at anchor without continuous electric power consumption. The Wolfson Unit MTIA in Southampton was commissioned to carry out model tests and a compromise was reached on the size of model, one to fifteen scale – big enough to give credible results but small enough to complete as a working model afterwards. One of the authors agreed to make the model hull to the original lines and spot-checked against the yacht itself as a very light glass/epoxy structure. Experimental results showed that a useful roll reduction was possible while keeping the bilge keel within the rather small envelope between the tight turn of bilge, the wall clearance and the drying out angle. On the other hand when underway the same reduction could be achieved by a modest alteration of course. The bilge keel idea was shelved but a useful education in ship behaviour was a bonus. The Wolfson Unit also provided a number of runs in the Southampton tank, towed from the carriage but not instrumented. The tank wavemakers were set to a typical southern North Sea wave spectrum, varying the wave heights from run to run. This was just the head sea condition but provided very clear indication when the yacht had had enough; beginning to slam and take large amounts of water over the bow. Although not quantified, this subjective view used to gauge weather forecasts has proved a very good safety measure in prudently planning voyages.

After the hull doubling, until recently the only original steelwork below the waterline open to the sea was the two tubes which surround the propeller shafts from where they exit the hull to the stern glands much further forward. In 2014 it became evident that one of these was reaching the end of its life, leaking slightly through pin holes. A major dry docking in 2015 saw the hull cut open and these tubes completely replaced. The work was arduous enough but simplified by finding that the tubes could be a direct replacement, being 4 inch heavy steel pipe with the same pipe thread. To further simplify the work, the new tubes were put in as two half lengths with a connector. This avoided disturbing the very substantial steel casting carrying the outboard end of the tube and incorporating a white metal lined bronze bearing.

6. SEA VOYAGES

Lilian ventured back to sea in 1988, fifty years after the last sea voyage (for which there was the engineer's logbook – mostly notable for complaints about the difficulty of starting and keeping the engines going). Fuel and lubricating oil supplies were a constant problem, as there were numerous oil leaks and incomplete combustion with a black exhaust plume that made *Lilian* look like a steamer. During and just after World War II the engines had been converted from blast injection to CAV Bosch solid injection. Many attempts to improve fuel economy, including servicing and modifying injectors and pumps short of a radical rebuild, were only marginally effective and it was assumed from old photographs that the blast injection originally fitted was more fuel-efficient. However, *Lilian* was rarely in a hurry, and although hull speed was never more than seven knots, that sedate progress was enough.

Most years since, *Lilian* has been to sea. The most notable voyage was to Stockholm in 1995, via the Kiel canal to Copenhagen and Gothenburg and over the Gota canal to Stockholm(for a Motor Yacht Society gathering), returning via the Baltic through Kiel again.Crossing the North Sea with the original engines revealed that they were very prone to overheating and salt precipitation in the direct seawater cooling system after continuous hard running for a day or two – they ran much cooler in fresh water or in the relatively brackish Baltic. Exhaust valves frequently stuck due to carbon accumulation, and engineers had to become adept at quickly replacing valve cages from an overhauled collection standing by.

Fortunately there was local engineering expertise at the Kew Bridge Engines Trust, a team keeping traditional beam engines and associated machines going with an extensive workshop. When a piston broke at sea, on a short, shake-down cruise a fortnight before attending a historic boats rally, the team made a pattern, cast a new piston, machined and re-installed it and sent *Lilian* on her way running it in faultlessly down Channel 500 miles to the rally at Brest. They also helped rebuild both gearboxes, white-metal and scrape in big-ends, and convert the engine starting system back from unsatisfactory later air motors to the original direct air start valves on three cylinders. They even drew up and commissioned a crankshaft when the port one broke on the way back from a journey to view the total solar eclipse mid-Channel in 1999.

Breaking a crankshaft could have spelt the end of the port engine, but fortunately minimal damage was done and over the past decade the engine has been gradually rebuilt and restored to working museum condition. For sixteen years *Lilian* has run with a modern port diesel engine, keeping the stern gear intact so the new gearbox had a 3:1 reduction to maintain 550 rpm at the propeller. Initially this was a Manta II 120hp 6cylinder marinised Ford, later changed to a JCB 4cylinder 85hp diesel to utilise hydraulics to run bow thruster and anchor winch. In 2011, the working starboard engine was extricated (first filming the intricacies of starting and running it), and replaced with a second JCB and hydraulics. Both old engines with gearboxes are in storage in working condition ready to be re-installed. Meanwhile the pair of new engines, which are cleaner, more powerful and less prone to breakdown, allow the boat to keep travelling, and the old engines are not risked by heavy usage. This avoids responsibility for a catastrophic damage event during an over-ambitious cruising plan.

A further deviation from originality in favour of practicality came with the fitting of a bow thruster. Because of the fine bow and limited draught forward, the tunnel diameter is technically undersized. However, it has proved an excellent addition. Turning the vessel around takes time but the thruster is very handy in holding the bow up waiting for locks in cross winds. The initial downside was that the highly rated electric motors cut out on overheat after a short time and had a long delay between going from port to starboard thrust and back. Burnt out motors was a factor in the decision to go to a hydraulic system as part of a more comprehensive upgrade.

Crews had become evidently weaker and less inclined to slave at the handles of the manual capstan/windlass. The hydraulic system, therefore, was sized to run both an anchor windlass and a hydraulic motor replacing the electric one on the bow thruster, which is now a reliable piece of equipment. This move also influenced the choice of modern main engines.

Although modern aids to navigation are available, a Walker's ship's log is still trailed, a sounding lead is used, and a sextant and a full set of signal flags are carried. Some of the original glassware and monogrammed Royal Copenhagen crockery is still extant. The only current lack is the original paid crew of eight (captain, engineer, steward, chef and four deck-hands).

7. TENDERS/SHIP'S BOATS

The two replica ship's boats were built to the original drawings by Richmond Bridge Boathouses (Mark Edwards) in 2000 (Figure 9). The carvel strip cedar hull construction was epoxy sheathed (and painted) to give strength and water-tightness, as these are swung from the original davits for use at sea. The Danish petrol engine was a relatively modern 12hp Vire, which could be removed and stored ashore in winter. The sailing rig was a small loose-footed lug, and there are three rowing positions (and a sculling notch in the stern) for flexibility. There have been entries in the Great River Race on the Thames as a 'historic ship's boat', winning the category twice. Both boats are quite comfortable and dry in a seaway, and are useful for running ashore or exploring when the yacht is anchored off while travelling.

Figure 9: The tenders constructed in 2000 to the original 1916 design

8. REFLECTIONS AND CONCLUSIONS

Vessels that reached their centenary have usually had a varied history, in this case too attractive or too useful to be scrapped [3]. In pulling a vessel back from derelict condition the main questions are what will the vessel be used for and how much money is available. In the case of *Lilian* the first goal was to make a comfortable family home on a residential mooring on the Thames. From there the goal moved to short cruises crewed by family and friends, then to longer and more ambitious voyages, so far in the range from Sweden to Brittany. Restoration and upgrading has been done out of a couple's post tax income over several decades without external grants or sponsorships. Over the years, there has been growing appreciation of *Lilian's* special place in maritime history, so the latest goal is preservation for future generations.

Opinions are strongly divided on doubling of steel hulls instead of letting in new material to replace wasted steel. In the case of *Lilian* doubling was well done and has undoubtedly given the old vessel a new lease of life – 34 years to date. However, the inner hull and frames have continued to deteriorate in places, so further hull work may be needed to enable cruising to continue in unsheltered waters.

In its hundredth year, the boat is in commission and goes to sea (Figure 10). *Lilian* travelled to Brest and Douarnenez in 2016,and was visited by grandchildren of the French owner who recognised the boat from family photograph albums from the 1920s (including a spell in the Mediterranean). It is hoped these voyages and discoveries will continue, as *Lilian* is still successfully doing what it was built for [4].

Figure 10: *Lilian* in 2016

9. ACKNOWLEDGEMENTS

We would like to thanks all those who have worked restoring and travelling on *Lilian*, especially Mrs Hilary Pereira, who also commented on this paper in draft.

10. REFERENCES

1. GFritz-Crone, Å Fritz-CroneCG Pettersson, Legendarisk Båtkonstruktor. SjøhistoriskaMuseet2003ISBN 91-85268-91-7

2. White R. Polar Z's 74 years in Lilian II. Marine Propulsion 1990 (May/June) 26-28

3. Prescott R et al.Conserving Historic Vessels.Understanding historic vessels vol 3.National Historic Ships 2010. ISBN 978-0-9566554-0-0

4. GotvedA, Berg A. for European Maritime Heritage Working Group 2002.The Barcelona Charter: European Charter for the conservation and restoration of traditional ships in operation.

11. AUTHORS' BIOGRAPHIES

Richard White was a seagoing engineer with Shaw Savill Line before going ashore to take an engineering degree. He joined the staff of Motor Ship magazine, and remains in marine technical journalism, having co-founded Marine Propulsion, edited Ship & Boat International for 13 years, the last eight after RINA acquired the journal, later started Offshore Support Journal. He is a Member of RINA and for many years has served on the Institution's Small Craft Committee.

Dr Scott Pereira is an immunologist and co-owner of *Lilian*, with his wife, Hilary. Previously he has owned narrow boat *Alcor*(built 1935) and *Sabrina*, a steam inspection launch (built 1870). He is currently honorary treasurer of the Maritime Heritage Trust.
[5745words]

This page intentionally left blank.

SALVAGE AT THE ENDS OF THE EARTH – The Ongoing Story of the Steam Trawler 'VIOLA'

R J H Spink, Solis Marine Consultants, UK
Dr R Robinson, University of Hull, UK
Captain J N D Simpson, Solis Marine Consultants, UK

SUMMARY

The former steam trawler 'Dias' (ex 'Viola') was built in 1906 at Beverley in East Yorkshire for the Hellyer Steam Fishing Company of Hull. Today, accompanied by other old whalers, she lies rusting at Grytviken, a short distance from the grave of Sir Ernest Shackleton, at the deserted whaling station on the remote South Atlantic island of South Georgia. In the next chapter of this vessel's tale a modern day salvage operation is about to commence in order to return her to Hull and, following an extensive restoration project, install her as a permanent memorial to the merchant seamen lost during the Great War.

Viola's story is unique: a remarkable story of fisheries, whaling, sealing, war and exploration during which she both weathered and witnessed many aspects of mankind's twentieth century struggles on the sea. Viola is now the oldest surviving former steam trawler in the world with her steam engines still intact. In the next chapter of this vessel's tale a modern-day salvage operation is about to commence in order to return her to Hull and, following an extensive restoration project, install her as a permanent memorial to the merchant seamen lost during the Great War.

1. INTRODUCTION

Built in Beverley in 1906 Viola currently lies in Grytviken Harbour, South Georgia. This iconic vessel is the world's oldest remaining steam trawler with steam engines still intact and, as a former requisitioned armed trawler, one of only four remaining British warships that saw active service in the First World War, spending four years on the maritime front line in the grim fight against mines and U-boats and participating in a substantial number of actions against enemy vessels, including the sinking of two U-boats. There is a strong attachment to Viola in Hull and area and a great deal of interest in coastal communities around the United Kingdom.

The Viola Trust has been established by a group of individuals interested in recovering Viola and returning her to Hull in order to preserve this historic vessel, to provide a permanent memorial to the merchant seamen that lost their lives in the First World War and to remember the seafarers of all nationalities who worked in some of the harshest conditions in the world to make a living.

The vessel is currently Crown Property, owned by the Government of South Georgia and the South Sandwich Islands (GSGSSI) which are amongst the United Kingdom's Overseas Territories. GSGSSI is aware of Hull's interest in the return of the vessel to the Humber and, in response, to a request from the Viola Trust backed by Alan Johnson MP, has indicated that they are minded to allow steps to be taken to refloat, recover and restore the vessel subject to appropriate conditions being met.

A preliminary feasibility study for the proposed salvage operation was prepared in 2015 and was followed up by a detailed site inspection in October 2016 with a view to finalising and preparing a tender document to be presented to salvage contractors interested in carrying out the operation. The findings of this recent survey are discussed later in this paper which commences with a brief overview of the history of Viola.

2. THE HISTORY OF VIOLA

2.1 THE TRAWLING YEARS

The steam trawler Viola was built for the Hellyer Steam Fishing Company in 1906. The Hellyer family, who originated in Devon, had owned a small fleet of sailing smacks in Hull at the end of the 1870s but during the 1880s they become involved in the construction of steam trawlers. Many of the steam trawlers built for Hellyers in the 1890s and early 1900s were intended for distant water trips to Iceland and the Faroes but by 1905 Charles Hellyer decided to build a completely new North Sea boxing fleet of steam trawlers. Hellyer's plan was staggeringly ambitious by contemporary fishing standards. He secured the necessary financial backing to construct a fleet of fifty vessels from scratch that would be serviced by five new steam cutters. This was probably the largest venture of its kind ever carried out by any fishing port in the world. The enterprise was reported to have required an investment of £320,000.

This ship was one of the trawlers built as part of this order. Constructed at Beverley by Cook, Welton and Gemmell, Viola was launched in February 1906 and then fitted with triple expansion steam engines constructed by Amos and Smith of Hull with an open bridge abaft a woodbine funnel. Like almost every vessel in the fleet Viola took the name of a Shakespearian character. Charles Hellyer had a fascination for the Bard, even naming his son Orlando. This Viola was 108 feet in length and 174 gross registered tons.

Down to the Great War, Viola worked as part of the Hellyer Boxing Fleet. This fleet of trawlers worked well out in the North Sea and transferred their fish on almost a daily basis to fast steam cutters that ran the catches into London's Billingsgate Market. The average length of a fishing trip for Viola and crew was around 37/38 days and they were worked hard. The vessel was regularly at sea for 310 days or more a year. Staggering though this may seem, the Hellyer fleet was probably continuously at sea from March 1910 until the outbreak of war in August 1914. As some trawlers returned to port when their fuel and food supplies ran low their places in the fleet were taken by newly replenished vessels: during this period there were thus always a large number of Hellyer vessels out on the North Sea grounds, regardless of weather or season. One of the poignant stories of its early years is that of Viola's sister ship, Antonio, built at the same time at Grovehill, Beverley. Antonio was lost without trace on the trawler's second voyage in 1906. This tragic loss is particularly ironic given the long career of the Viola.

2.2 THE FIRST WORLD WAR

Shortly, after the outbreak of the Great War Viola was requisitioned by the Admiralty, armed with a 3-pounder gun and assigned to Auxiliary Patrol Area 2 based on Shetland, working on patrol operations with occasional boom duties and escort work. Patrol duties often took Viola from Muckle Flugga to Fair Isle. Whilst stationed in Shetland Viola saw action against at least one U-boat and one member of her crew also drowned and is buried in a military grave in Lerwick. The ship's crew were all fishermen who had joined the Royal Naval Reserve and mainly hailed from Hull and Shetland. Mary Allum, the wife of Charles Allum (Viola's skipper at the time), was one of a number of fishermen's wives who followed their husbands to war and went with her children to live in Lerwick, where she unfortunately died of corrosive sublimate poisoning in September 1916, possibly whilst treating sphagnum moss which was being gathered by local women for use as field dressings on the Western Front. At that time field dressings were rapidly being depleted on the Western Front and this was a novel and effective way of dealing with the problem. The Allum children were initially fostered in Shetland and well looked after by the local community. This issue of women from fishing communities, their support for the fishermen fighting the war at sea and their role in the gathering of sphagnum moss are aspects of the war which have been largely forgotten.

By September 1916 the vessel had been refitted with a 6-pounder gun and, shortly afterwards, transferred to the River Tyne for anti-submarine patrol operations. Depth charges were by then being deployed and Viola was soon equipped with this new weapon. The armed trawler was also fitted with hydrophones, an early electronic listening device. In 1917 Viola opened fire on a U-boat attacking a merchant ship off the Farne Islands and drove it off after an exchange of gunfire. That year it was also involved in the rescue of the crew of a French coal barge which was being driven ashore at Scarborough in a fierce storm. Viola's skipper, Charles Allum, was mentioned in despatches for these events. The following year the vessel was fitted with a 12 pounder gun.

On the 13th August 1918 Viola, together with the armed trawlers John Gilman and John Brooker, played a major role in the sinking of the UB-30 off Whitby. Late in September 1918 Viola was again involved with other armed trawlers, destroyers and an airship, R27, in the destruction of the UB-115 off the Northumberland coast. This action is believed to have been the first to have involved the use of an airship in the destruction of a submarine and the UB-115 was probably the last U-boat sunk in conflict in the North Sea during the Great War.

Charles Allum, from Hessle Road in Hull, was Viola's skipper for much of the war. The vessel spent over four years on the maritime front-line, part of the grim war to keep the sea-lanes around the British Isles and beyond open, a seemingly never ending conflict fought against mines and U-boats. Day after day, week after week, month after month this armed trawler was on patrol and probably steamed far more miles in the Great War than any single dreadnought. During the Great War at least fifteen trawlers from the Hellyer fleet of fifty vessels were lost to enemy action whilst fishing or on War Service.

2.3 FISHING AND WHALING IN NORWAY AND AFRICA

Viola was decommissioned in Milford Haven early in 1919 and returned for fishing. Hellyers, however, had by this time decided to concentrate on distant water fishing off Iceland, Barents Sea and the like and had sold off their remaining North Sea fleet. Viola was eventually purchased by Norwegian owners, renamed Kapduen and was one of the first trawlers to work out of Norway. After a few years, however, the firm that owned her was taken over by Nils Torvald Nielsen Alonso a veteran whaler. The trawler was converted into a whale catcher with a new bridge forward of the funnel and was involved in several major whaling trips over the next few years off the coast of Africa. The first of these, with the factory ship Bas II was quite historic, being the first expedition to carry out non-licensed pelagic whaling off the coast of Africa. None these expeditions proved a real commercial success and by 1927 the former trawler, now renamed Dias, was laid up in Sandefjord in Norway.

2.4 SAILING AND EXPEDITION WORK IN THE SOUTH ATLANTIC

In 1927 the vessel was sold to Compania Argentina de Pesca Sociedad Anonima, known as Pesca for short.

Pesca operated from Grytviken in South Georgia and required a vessel for sealing. The labour force and operational management of the company were mainly Norwegian. Viola/Dias voyaged to Grytviken that year to take up duties. Henceforward, the former trawler was used for hunting elephant seals. This activity was very carefully regulated by the Falkland Islands Government and is an early example of successful conservation – in contrast to whaling. The island was split into divisions, one area of which was rested each year. Only adult bulls could be taken and a close season was introduced.

Since Dias had substantial cargo carrying capacity, the vessel was also in demand to support expeditions both to South Georgia and the South Atlantic. These included the relief of the Argentine meteorological station at Laurie Island. The round trip from South Georgia to the South Orkneys took around twelve days but was depended on ice conditions. On one occasion an earlier attempt to relieve the weather station with an Argentinean vessel had failed but the ex-trawler succeeded in getting through. In some years the little vessel would sail to Buenos Aries to take on board expedition personnel. Dias also assisted in various expeditions to South Georgia. One of the earliest was the Kohl-Larsen Expedition of 1928/9 which took the first cinematographic film of the island. Others included the British South Georgia Expedition under the leadership of the climber George Sutton, biological work carried out by the Falkland Islands Government, the Bird Island Expedition of 1958 and topographical surveys carried out by Duncan Carse.

Such useage over many decades must make Viola-Dias one of the longest serving vessels to be involved in South Atlantic expeditions. Around the time of the Kohl-Larsen expedition the then Swedish skipper, Johan Johannesson, located a forgotten harbour bay whilst on a sealing expedition to the inhospitable south west coast. Here, he found the remains of an old sealing camp. The bay was subsequently named Dias Cove.

2.5 LAY UP IN GRYTVIKEN AND THE FALKLANDS WAR

In the 1950s the ship's steam engines were converted to burn oil instead of coal. In 1960 Pesca sold out to the British firm Albion Star and the ex-trawler returned to the British flag. In 1964/5 the whaling station of Grytviken was closed and Dias, together with the other surviving vessels Petrel and Albatros, was mothballed and laid up. A caretaker looked after the little fleet until 1970. During the 1970s, left unattended, the vessels, settled in the water under the weight of accumulated winter snow.

The ex-trawler, although laid up, had an unwitting role in the start of the Falklands War. The crisis, of course, came to a head when a group of Argentinian scrap metal merchants landed on South Georgia. Amongst their agreed tasks was the cutting up of the three vessels, which included Dias, laid up at Grytviken.

During the Falklands War in 1982, South Georgia was, of course, briefly occupied by Argentina and British forces re-took the island in an action fought in the neighbourhood of the old trawler. As a former requisitioned minesweeper, Dias must be the only vessel to have 'seen action' in both the Great War and the Falklands War. The action to re-take the island involved the disabling of the Argentinean submarine, Santa Fe in the waters just in front of the ex-trawler. This was the first time helicopters had taken a submarine out in war and a remarkable reminder of the fact that the Viola had been involved in the first action in which an airship had been involved in the sinking of a U-boat in September 1918. Shortly after the Argentine capitulation, five Hull trawlers requisitioned by the Admiralty, arrived in Cumberland Bay outside Grytviken. They were the last Hull trawlers to be requisitioned by the Admiralty and they were moored quite close to where Viola-Dias, as one of the first trawlers requisitioned – almost seventy years earlier – was resting. In the haste of war no one noticed the coincidence at the time.

2.6 OIL REMOVAL WORK

In the later 1990s a plan to take the vessel to the National Fishing Heritage Centre at Grimsby was formulated. A marine survey was carried out which concluded that it was possible to return her to England and restore her to working order. It also noted that many spare parts were available in Grytviken's engineering stores. Unfortunately, this was not proceeded with.

By the early 2000s there was concern over the leakage of fuel oil from the Dias and other vessels laid up at Grytviken. Remedial work took place in 2002/3 and over the first weeks of 2004 there was considerable activity around the vessels. A team has carried out further work to remove the oil from Dias and Albatros. At the end of January Viola/Dias was floated for the first time in years. The Dias and Albatros have since been secured in an area dug out of the beach in front of her former berth.

Whilst the vessels at Grytviken were secured for a time, their long-term future remains uncertain.

2.7 THE VIOLA TRUST

In recent years, further attempts to recover Viola-Dias have been mooted. In late 2014 a survey of the vessel was organised and funded by a group of Hull companies led by Paul Escreet of SMS Towage. In 2015 an initial feasibility report was also produced by Solis Marine Consultants and the positive outcome of the study has prompted the creation of the Viola Trust to recover the ship, arrest its deterioration and return it to its home port. Members of the Viola Trust bring their particular skills to this unique project. The Trust is headed by Paul Escreet, chairman of SMS Towage, the UK's largest independent towing company. Other members include, Dominic Ward, Senior Partner at the law firm Andrew Jackson who has a vast experience in the field of maritime law, retired Admiral Nick Lambert, former captain of HMS Endurance and Dr Robb Robinson, maritime historian at the University of Hull. Patron of the Trust is Alan Johnson, MP and the organisation is supported by various individuals and companies who are lending their experience and support.

The history of this iconic vessel is quite a story in itself. Viola/Dias is a unique survivor, virtually the last of the North Sea fleeters and probably the oldest surviving steam trawler with steam engines still intact anywhere in the world; a veteran of the Great War as well as a range of activities in the South Atlantic. This ship is truly a historic representative of Kingston upon Hull's unique maritime heritage and pioneering spirit and is also reminder of the remarkable role played by over 3000 fishing vessels and their crews, drawn from coastal communities across the British Isles and taken into Admiralty services to fight the grim war against mines and U-boats which was waged around the coasts for over four years.

In September 1914 Viola sailed off to war with a crew of fishermen, fought for more than four years on the maritime front-line and has yet to return to its home port from that Great War voyage.

3. VESSEL DESIGN AND LAYOUT

Very little information relating to the design and construction of Viola is available due largely to the age of the vessel. A copy of a basic general arrangement plan for the Falstaff class of steam trawlers is available although the quality is poor and copying over the years has warped the drawing to the point where reliable measurements can no longer be lifted from the plan. This plan is the only available documentation of the vessel's design discovered to date and there is no lines plan or even body section drawings of the vessel at all. A digitised version of the general arrangement is provided at Figure 1 and the principal particulars taken from the same plan are detailed below.

LBP:	108 feet 6 inches	(33.1 metres)
LOA:	120 feet 9inches	(36.8 metres)
B:	21 feet 6 inches	(6.5 metres)
Depth of Hold:	11 feet 6 inches	(3.5 metres)

The vessel has five main compartments separated by transverse bulkheads. The aftermost compartment, above the shaft space, was used as accommodation for the master and engineer and forward of that is the engine and boiler room. The third compartment is the fish/cargo hold, forward of which is further crew accommodation area and the fore peak with associated chain locker and store.

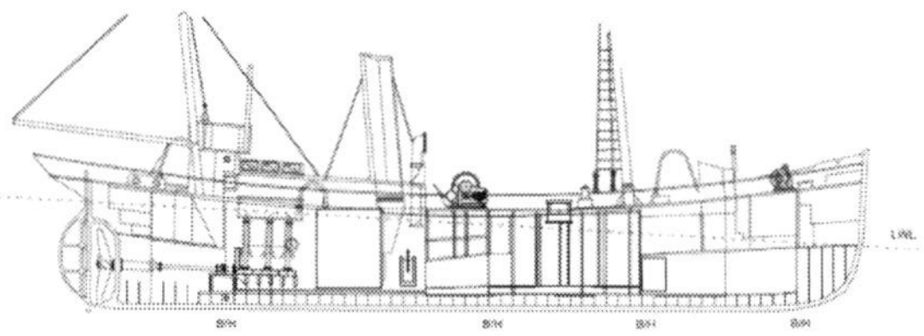

Figure 1 Digitised version of general arrangement plan

The vessel is of riveted construction and has a flat bar keel running the full length of the centreline which protrudes slightly below the bottom of the hull plating. Transverse frames provide the main structural support as there are no longitudinal members other than the keel and two bilge keels. The deck is of wooden construction rather than steel and is laid over the transverse frames and made watertight through caulking. There are two bunker tanks that were originally designed for carrying coal and were converted for fuel oil in the 1960s, but other than those there are no other tanks within the vessel unless there is a small fresh water tank in one of the flooded compartments.

Viola currently lies in a trench dug out of the beach during the 2004 oil removal operations next to the whale catcher Albatros at their former berth alongside the Harpon Jetty in Grytviken, as shown in Figure 2 accompanied by one of the many elephant seals that inhabit the same area. All compartments are flooded and she is firmly aground.

Figure 2 Viola in current position in Grytviken, South Georgia

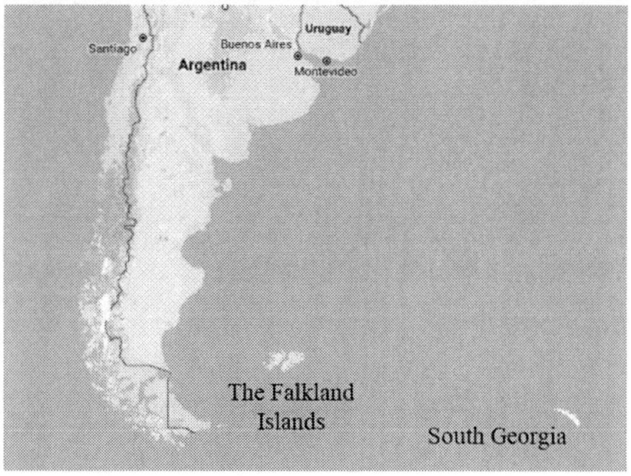

Figure 3 South Georgia and the Falkland Islands

4. PLANNING FOR THE JOURNEY HOME

The operation to return Viola to her home port of Hull will consist of two phases; firstly the patching, dewatering and refloating work in Grytviken, which would also include a period of stabilisation and potentially further pumping and / or patching, and secondly the transportation back to Hull in the UK. South Georgia is approximately 8,300 nautical miles from Hull by sea.

4.1 STAGE 1 – REFLOATING

The feasibility study carried out by Solis Marine Consultants in 2015 identified a number of options for the recovery and transportation of Viola back to the UK. The refloat operation in Grytviken would utilise standard salvage techniques for the restoration of buoyancy including patching and pumping with the possible use of air bags for additional buoyancy if required.

The salvage team will consist of a salvage master, salvage engineer and at least one team of divers and riggers. A supply vessel will be required to transport the team and equipment to South Georgia as there is no airport; it will be essential that the operation is entirely self-sufficient since resources on the island are very limited and the closest port is Stanley in the Falkland Islands which are some 800 miles away by sea. Planning and preparation are key to any modern day salvage operation but in this case the omission of the most basic spare part could directly affect the overall success of the project.

4.2 STAGE 2 – TRANSPORTATION TO UK

The second phase is the transportation back to the UK, which probably presents the most challenging aspect of the project. Since the vessel is unlikely to survive a wet tow the preferred option for transportation is by either by project cargo vessel or semi-submersible heavy lift ship. This will require construction of a bespoke lifting cradle onto which Viola would be floated to then be lifted onto the transportation vessel. The design of the cradle in itself presents a challenge given the fact that the exact dimensions and shape of the hull are not known.

If a project cargo vessel is used then determining the weight of Viola becomes critical since the transportation vessel must have sufficient capacity to lift her, along with the cradle, onto the deck. Such a vessel would also be operating on a fixed schedule, albeit on a detour from a regular route, and as such, any delays in the preparation work on Viola could potentially result in quite literally 'missing the boat'.

A semi-submersible vessel would still require a cradle with adequate sea fastenings although the weight aspect becomes less critical since such a vessel would have more than sufficient capacity for a vessel the size of Viola. Timing may also be more flexible since the vessel would likely be chartered specifically for this project. The main drawback with this approach would be the availability and cost of the transportation vessel, the latter of which is likely to be prohibitive.

Despite the challenges identified the feasibility study concluded that the recovery project is a viable one and this prompted the creation of The Viola Trust which will coordinate the recovery of Viola, the return to Hull and then its restoration before taking its place as a permanent memorial.

4.3 THE 2016 SURVEY

Following the establishment of The Viola Trust in September 2016, Solis Marine Consultants carried out a detailed site inspection in South Georgia with a view to allowing a final decision on feasibility and the production of a tender document to go out to salvage contractors interested in being involved and taking the project to completion.

The main objectives of the survey were to gather as much information as possible in relation to the shape of the hull form, the general condition of structural members and her current position, draughts, trim and list. Additional work investigated the potential effect of the removal of Viola on the neighbouring vessel, Albatros, to which she is currently moored against as well as assessing various local considerations in relation to environmental protection and the logistical issues when working at such a remote location.

The survey successfully confirmed the feasibility of the recovery operation and sufficient information was collated to allow the preparation of a salvage plan with supporting pollution prevention and HSE plans as well as an environmental impact assessment.

5. HYDROSTATIC MODELLING

5.1 DEFINING THE HULLFORM

Hull offset measurements of Viola were taken for the areas above the water line with a view to developing a hydrostatic model of the vessel that would assist primarily in providing a weight estimate.

The trawler 'Sheraton' was built at the Cook, Welton and Gemmell yard around the same time as Viola and although differing in overall dimensions, was built to a very similar design. The wreck of Sheraton, consisting only of the lower sections of the hull, is situated on the beach at Hunstanton, Norfolk and is accessible at low tide. Measurements of the remaining structure have been recorded and scaled in order to build up a picture of the form of Viola below the waterline. This information, along with the measurements taken from Viola itself has been used to create a hydrostatic model, shown in Figure 4, which will assist throughout the recovery project.

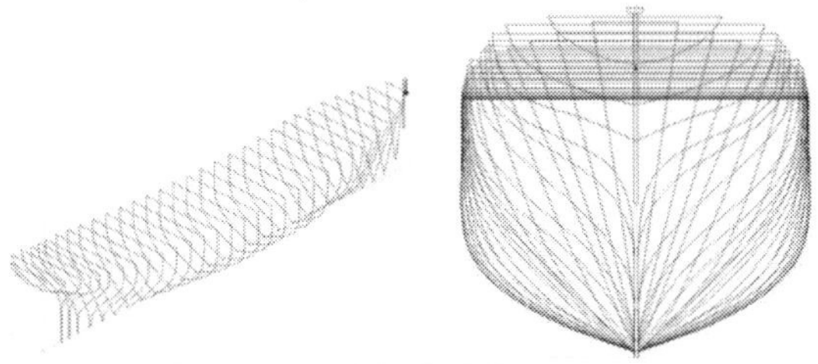

Figure 4 Hydrostatic model of Viola

5.2 WEIGHT ESTIMATE

During the 2004 oil removal operation, Viola was refloated for a period of around six weeks; photographs from this time have been used to estimate her floating draughts which have been applied to the hydrostatic model. This has provided an initial weight estimate of around 335 tonnes.

6. ENVIRONMENTAL CONSIDERATIONS

6.1 WILDLIFE AND HABITATS

Grytviken is home to a wide variety of wildlife including a number of highly sensitive species. As with all modern day salvage operations, the protection of the environment is accorded the highest priority of the recovery operation after the safety of those working on the project.

The primary concern in relation to potential disturbance of wildlife is likely to be the fur and elephant seals, which come ashore to give birth to their pups around December / January each year and then stay ashore with their young through the summer months. During this time, large numbers of seals occupy the waters and shore line in the vicinity of Viola as well as the Harpon Jetty.

As well as protecting the natural habitat of these animals during their breeding season it must also be remembered that they can be dangerous to humans if disturbed during this time. The general approach of the salvage team will therefore be to give way to the seals, maintain a safe distance of at least five metres from them and cease work if necessary.

6.2 BIOSECURITY

Following significant investment into the eradication of non-native species such as rodents, reindeer and invasive weeds from South Georgia there are stringent biosecurity measures in place to prevent any non-native species from entering the territory. These measures are laid out in the Government of South Georgia and the South Sandwich Islands Biosecurity Handbook 2016-2017 and all vessels involved in the recovery project will have to comply with these requirements.

6.3 ASBESTOS

An asbestos survey of 'VIOLA' was carried out in 2012 by Thomas Moore and as a result, asbestos is known to be present in the boiler room, steering gear flat and the engine room, where it was used as insulation on the boiler casing, the engine block and various pipework. The asbestos insulation is in poor condition and therefore must not be disturbed until it can be removed by a licensed contractor. Asbestos is also present in pipe flange gaskets throughout the vessel and although this is in reasonable condition it should also not be disturbed.

It is not intended that any asbestos will be removed in South Georgia. The requirement will be to prevent the material becoming airborne (friable) until the spaces can be sealed for the remainder of operations and transportation.

Access to the engine room and boiler room will be restricted during salvage operations with only limited work to be allowed. Where essential work is required in these spaces, pumping out water / sediment for example, then this would be carried out at a safe distance from the asbestos so as to prevent disturbance and personnel would be trained in asbestos awareness and equipped with suitable PPE.

When pumping out the space, a filtration system will be in use and once the area is accessible all asbestos sites will sprayed with a PVA solution in order to seal the fibres. This will minimise any friable material escaping the confines of the vessel. Following the securing of the space it will be sealed and suitably labelled until arrival in the UK where asbestos removal experts will take over to clear the material in compliance with UK regulations. The guidance of the UK HSE and the relevant publications will form the baseline standard for all operations.

6.4 POLLUTION PREVENTION

As with any salvage operation, a detailed pollution prevention and response plan will be prepared in support of the salvage plan itself. This document, usually prepared by the contractor, details the equipment to be provided on site, measures to be taken during operations to contain any release of pollutants and procedures for responding to a release. The plan will be prepared in conjunction with the other documents detailing the management and logistics of the project, all of which must meet with approval from the relevant authorities, namely the Government of South Georgia and the South Sandwich Islands.

6.5 SITE REMEDIATION

The salvage plan will include due consideration to remediation works at the site so as to ensure it is returned to its original condition, albeit in the absence of Viola. This remediation work is likely to consist of the replacement of seabed material in order to fill the trench that the vessel currently rests in as well as the restoration of any disturbance to the shore from the use of excavating equipment. There should be no visible long term effects on the site following removal of Viola and it will therefore remain safe for wildlife and passengers.

7. SOUTH GEORGIA HERITAGE

Viola, or Dias as she is known in South Georgia, spent a large proportion of her working life in the South Atlantic and her presence in Grytviken is of considerable significance to those with both personal and professional links to the heritage of South Georgia and the whaling industry. The importance of her time there should not be underestimated and, whilst it is not feasible for her to be restored and maintained there, it is important that her influence over the course of history in South Georgia can be suitably reflected both in Grytviken and back in the UK.

It has been suggested that a brass plaque or memorial based on the history of Viola could be installed at the Harpon Jetty following the removal of the vessel. The intention would be to provide a tribute to the men, of all nationalities, that sailed on Viola / Dias during her lifetime. Further to this, a book or film describing the recovery project could also be produced for display at the museum at Grytviken as a means of providing further information for visitors to South Georgia.

8. CONCLUSIONS

This project presents a unique combination of historic significance within the context of a modern day salvage operation. The history of Viola is already an intriguing and exciting tale of trawling, whaling, sealing, war and exploration affecting the lives of hundreds of merchant seamen over the course of a century and longer. This next chapter in Viola's story sees her as the subject of a recovery operation utilising the latest technology and specialist expertise in the salvage and shipping industries to return the vessel to her home port.

The salvage itself is unique and presents challenges not encountered in most commercial salvage operations; breeding seals being one of those. The coordination and planning required to carry out an operation at such a remote location is complex but can be compared with other operations in far flung corners of the world. However, in the context of a charitable project there is an added dimension relating to the of securing the funds for such a venture and so the challenge for this vessel continues.

9. AUTHORS BIOGRAPHY

Rosalind Spink holds the current position of Partner at Solis Marine Consultants and is a member of the Panel of Special Casualty Representatives (SCR) at Lloyds. She is a naval architect working as a consultant in relation to shipping incident investigation, salvage and wreck removal. Her previous experience covers a number of complex and politically sensitive salvage and wreck removal operations worldwide.

Dr Robb Robinson holds the current position of Research Associate and Lecturer at the University of Hull and is a Trustee of the Viola Trust. He is also a Trustee of the British Commission for Maritime History, Treasurer of the North Atlantic Fisheries History Association, and worked with colleagues from various countries on the production of a two volume History of the North Atlantic Fisheries. He also has an interest in the Great War at sea from the perspective of the smaller vessels involved and also the island of South Georgia. His teaching interests include National and International Fisheries History, Whaling and the Maritime History of the Yorkshire Coast and the North Sea.

Captain John Simpson also holds the current position of Partner at Solis Marine Consultants and is also a member of the Panel of Special Casualty Representatives (SCR) at Lloyds. He is a master mariner working as a consultant in the field of maritime incidents. His previous experience includes time at sea as master of various types of vessels and as a pilot on the River Humber. John has worked on many salvage and wreck operations in such remote locations as the Azores.

This page intentionally left blank.

RMS MAURETANIA – DESIGN AND SIMULATION OF CUNARD'S 'ATLANTIC GREYHOUND'

K W Hutchinson, Babcock International Group - Energy and Marine Technology, UK
M Irving, South Tyneside College – South Shields Marine School, UK
G Horseman, Siemens Power Generation Services, UK

SUMMARY

This paper discusses the background and technical developments regarding ship design and propulsion directly preceding and influencing the design of the steam turbine propelled trans-Atlantic liner *RMS Mauretania*. Over a century after her construction at Wallsend on the River Tyne, the modelling on the Ship Simulator at South Shields Marine School, also on the River Tyne, is discussed and illustrated. Unfortunately, but understandably, *Mauretania* is no longer operational as she was broken up in 1935. However, the *Mauretania* now exists again as an operational ship, albeit virtually, on the state of the art Ship Simulator at South Shields Marine School. The mathematical modelling and detailed graphical representation of the *Mauretania* are discussed. The 'virtual' *RMS Mauretania* is presented in order to demonstrate the realism of current simulation and, of course, the performance of this truly iconic and record breaking ship, which will hopefully encourage others as to the benefits of the realistic simulation of historic ships and hence 'bringing them back to life' for the current and future generations.

1. INTRODUCTION

This paper aims to briefly examine the history, design and simulation of the groundbreaking and record-breaking Edwardian 'Greyhound of the Atlantic' *RMS Mauretania* which was built 110 years ago. Figure 1 below is a representation of this event painted by John C Currie, a Naval Architect with Babcock on Tyneside who worked at (the original) Swan Hunter Shipbuilders until it's closure in 1995. This painting was commissioned by Swan Hunter Shipbuilders in 1991 and prints were presented to the Launch Party on the occasion of the launch of the Royal Navy's Type 23 Frigate *HMS Richmond* at the Wallsend Yard on 6th April 1993, she finally sailed from Wallsend in early November 1994. It is interesting to note that, Ship 137 as she was known, was the last ship dynamically launched on the River Tyne and she was built on Number 2 Berth at the Wallsend Yard which was in the same approximate location as the covered Berths built 90 years earlier specifically for the construction of *RMS Mauretania*.

Figure 1: *RMS Mauretania* leaving the River Tyne 22nd October 1907

RMS Mauretania, together with her slightly older sister *RMS Lusitania*, were the first of the giant trans-Atlantic passenger liners designed with steam turbine propulsion with the intention of winning the Blue Riband. The Blue Riband is an unofficial prize awarded to a passenger liner in regular service which has made the fastest westbound crossing of the Atlantic based on an average speed. The competition began in 1838 although there was not an actual trophy until Harold Hales, a British politician and shipping owner, donated the Hales Trophy in 1935.

These two sisters were conceived in 1902 when the Cunard Steamship Company decided to build two gigantic liners to win the prize, and Sir Charles Parsons proposed that the liners should be powered by his steam turbines rather than steam reciprocating engines. As the advantages of steam turbines far outweighed the disadvantages for a trans-Atlantic liner, in 1904 the original specification of steam reciprocating engines was changed to steam turbines. The *Lusitania* and *Mauretania* were the only ships with direct-drive steam turbines to hold the Blue Riband, as later holders all utilised a reduction-geared steam turbines. This direct-drive configuration, used on all the giant passenger liners up to the mid 1920s, was the same as that utilised a decade earlier in *Mauretania's* forebear the 1894 Tyne built 103 feet long experimental Steam Yacht *Turbinia*, designed by Sir Charles Parsons to demonstrate the superiority of his steam turbine for ship propulsion. *SY Turbinia* was the subject of a paper [1] by the principal author to the previous Historic Ships conference held by the Royal Institution of Naval Architects in late 2014, and hence this paper will recap and progress the story of steam turbine propulsion onwards into the first decade of the twentieth century through naval warships to the largest and most prestigious ships of their time, the trans-Atlantic liners.

2. SIR CHARLES ALGERNON PARSONS (1854 TO 1931)

The relationship between *RMS Mauretania* and *SY Turbinia* is obviously neither in size nor purpose, rather in the revolutionary prime movers they shared, namely the Parsons Marine Steam Turbine designed by the great Tyneside engineer Sir Charles Algernon Parsons [2]. The rapid development of Steam Turbine propulsion of ships is testimony to his foresight and perseverance. Sir Charles, a mathematician by education and a mechanical engineer by training, having served an apprenticeship with W. C. Armstrong and Co. in Newcastle-upon-Tyne, had, until 1893, concentrated his efforts on land based enterprises, principally electrical power generation using his patented Axial Flow Steam Turbines. Until then his interest in ships and the sea could only have come from his rowing prowess at Cambridge and his holidays on the reciprocating engine powered Steam Yacht *Titania* belonging to his father, the 3rd Earl of Rosse.

Sir Charles combined his interest in Steam Turbine technology with his passion for the sea and ships to design, construct, develop and demonstrate, in the remarkably short period of just five years, a prototype ship that was a first in terms of prime mover power source and propulsor technology and which set a record breaking confirmed trial speed of 32.76 knots. The ship in question is of course the first ever turbine steamer *SY Turbinia* which was built on the River Tyne in 1894. As discussed in Section 3, the impact of this little ship, when demonstrated three years later, resulted in the rapid adoption of steam turbine propulsion in both warships and passenger ships. As illustrated in Sections 4 and 5, the developed power of steam turbines quickly increased leading to greatly improved speeds. The size of steam turbines grew at a truly amazing pace, from approximately 2,000 shp (1,491 kW) in *Turbinia* of 1897 to in excess of 23,000 shp (17,151 kW) in *HMS Dreadnought* of 1906. This rapid development, after only a decade, culminated in the staggering 76,000 shp (56,673 kW) steam turbine set installed on the famous Tyne built Blue Riband holding leviathan *RMS Mauretania* – the topic of this paper.

3. *STEAM YACHT TURBINIA* (1894 TO DATE)

SY Turbinia is arguably one of the most famous and influential ships in the world as she set the standard for the next generation of steamships that directly followed her, the majority of which were subsequently turbine powered. *SY Turbinia* was designed as an experimental ship by Sir Charles Parsons in order to demonstrate the superiority of his axial flow steam turbine for ship propulsion. She was 103 feet 3 inches (31.471) long by 9 feet (2.743 metres) beam, had a 3 feet 4.5 inch (1.023 metres) draught displacing 42 tons (43 tonnes). She was built by Brown and Hood in Wallsend on the north bank of the River Tyne and launched in 1894. Until 1899 she was the fastest ship in the world, reaching an unofficial peak speed of 34.6 knots from three axial flow steam turbines delivering a total of about 1,975 shp to 2,200 shp (1,473 kW to 1,641 kW) at around 2,200 rpm, each turbine directly connected to a propeller shaft fitted with three equally sized three bladed propellers giving a total of nine propellers! This arrangement was arrived at after Sir Charles had encountered, investigated and overcome cavitation, and in the process, developing and constructing the world's first Cavitation Tunnel.

Turbinia's ground-breaking performance was demonstrated dramatically when she arrived, unannounced but not unexpected, at the Navy Review for Queen Victoria's Diamond Jubilee at Spithead on 26th June 1897 – Engineer-in-Chief of the Fleet Sir John Durston had proposed such a demonstration. In attendance were the Prince of Wales, Lords of the Admiralty, Members of Parliament, foreign dignitaries such as premiers and ambassadors and including Prince Henry of Prussia, the brother of Kaiser Wilhelm II, and of course the world's press. She famously raced at 34.5 knots between the four assembled lines of over 150 warships extending over 25 miles, as can be seen in Figure 2 below, and steamed up and down in front of the crowd evading with ease a Royal Navy Picket Boat, that endeavoured to pursue her, and almost swamping it with her wake!

Figure 2: *SY Turbinia* at the Spithead Review 1897 [3]

The design and performance of the *Turbinia* are discussed in [1] and [4].

Following *Turbinia's* performance at the 1897 Spithead Review and further trials attended by the Admiralty, as discussed below in Section 4, the Royal Navy ordered its first steam turbine propelled Torpedo Boat Destroyer in early 1898. *Turbinia* steamed to Paris in 1900 but by this time her job was done as the marine steam turbine was accepted by the shipping world as, as discussed in Section 5, the first steam turbine propelled merchant ship was laid down that year.

Tragedy, literally, struck on 11th January 1907 when *Turbinia* was nearly cut in half by the *SS Crosby* when it was launched from Hebburn-on-Tyne on the south bank opposite Parsons' works. However, even given the severe damage she was repaired in time to be pictured alongside the *RMS Mauretania* prior to her sailing (see Figure 20 later in the paper). However, mechanical problems prevented *Turbinia* from steaming with *Mauretania* down the River Tyne at the beginning of her delivery voyage on 22nd October 1907 – her absence can be seen in Figure 23 showing *Mauretania* leaving the Tyne for her unofficial builder's sea trials on 17th September 1907 but has now been rectified through simulation as shown in Figure 24 and other examples of the simulations performed.

In 1908, only 14 years after her launch, *Turbinia* was lifted out of the water for the last time in an attempt to slow down further deterioration. In 1926 she was offered to the Science Museum in London, but as the museum did not have the space to accommodate a ship of over 100 feet long the decision was made to cut her in two and hence only the aft end, containing the turbines and propellers, was transported by road to London with the remainder staying on Tyneside. In 1944 the fore end was presented to Newcastle Corporation and placed on display in Exhibition Park. In 1961 the two halves of *Turbinia* were finally reunited, the Science Museum taking the aft end off display in 1959. On 30th October 1994, 100 years after her launch, *Turbinia* was moved by road to Museum of Science and Engineering, now the Discovery Museum, in the centre of Newcastle upon Tyne where she has pride of place in the atrium of the museum, as can be seen in Figure 3.

Figure 3: *SY Turbinia* as currently displayed in the atrium of Newcastle-upon-Tyne's Discovery Museum [Tyne and Wear Archives and Museums]

4. STEAM TURBINE PROPELLED ROYAL NAVY SHIPS (1898 TO 1907)

In 1895 the French Torpedo Boat *Forban* had achieved a speed of 31 knots and no British Destroyers of the period could match such a speed. Hence, Sir Charles Parsons proposed building two Torpedo Boat Destroyers powered by his steam turbines rather than steam reciprocating engines.

4.1 *HMS VIPER* (1900)

On 4th March 1898, the Royal Navy ordered a three funnelled 210 feet 3.5 inches (64.099 metres) long by 21 feet (6.401 metres) beam, displacing 344 tons / 395 tons (350 tonnes normal / 399 tonnes deep) Torpedo Boat Destroyer *HMS Viper* from Parsons Marine Steam Turbine Company. As with the *Turbinia*, the construction of the hull had to be subcontracted, hence the ship was built by R. & W. Hawthorn Leslie and Company. She was launched from their Hebburn-on-Tyne shipyard on the south bank of the Tyne on 6th September 1899 and commissioned in 1900. The turbines produced 10,600 shp (7,904 kW) driving four shafts directly at 1,000 rpm with two propellers per shaft. She achieved 12,300 shp and 33.57 knots on trials, comfortably in excess of contract speed of 31 knots. The two high-pressure, HP, turbines drove the outer shafts and two low-pressure, LP, turbines (and astern turbines) the inner shafts. This is the exact configuration that was used on *RMS Mauretania* later that decade.

4.2 *HMS COBRA* (1900)

Armstrong Whitworth built, on speculation, as Yard Number 674, a four funnelled 223 feet (67.970 metres) long, 400 tons (410 tonnes) Torpedo Boat Destroyer with 11,500 shp (8,576 kW) Parsons turbines driving directly four shafts. She was launched on 28th June 1899 from their Elswick shipyard on the north bank of the River Tyne upriver of Newcastle. She was subsequently purchased by the Royal Navy on 8th May 1900 and named *HMS Cobra*. The ship had her first steam trials in June 1900 and was completed by September 1901 achieving 34.6 knots.

4.3 INITIAL FINDINGS

Tragically, both ships were lost in 1901 but not after convincing the Admiralty as to the merits of steam turbine propulsion. The advantages over steam reciprocating engines for naval ships were seen to be:

- powerful hence fast ships – a great advantage for a warship, and full power could be sustained;
- almost no vibration, hence less damage to the ship or crew fatigue;
- small and light, hence increased carrying capacity and low down in the ship so less vulnerable to hostile fire;
- more efficient at high / full power / speeds, hence lower running costs.
- more reliable.

However, there were disadvantages:

- large number of Stokers were needed and hence maximum speeds could only be maintained for short periods, similar to reciprocating engines – this was addressed in 1907 with the first use of

oil, and by 1913 oil had become the main fuel in the Royal Navy;
- less efficient at lower (cruising) powers / speeds – hence, turbine efficiencies had to be improved at lower power outputs;
- due to direct drive, turbines and propellers ran at non-optimal speeds – use of gearing to allow the turbine to run at high speeds and the propellers at low speeds, first used in a warship in 1911.

4.4 *HMS VELOX* (1902)

Following on from the success of *HMS Viper*, Parsons Marine Steam Turbine Company built on speculation another three funnelled Torpedo Boat Destroyer named *Python*. She was 215 feet / 210 feet 0.03 inches (65.532 metres overall / 64.018 metres between perpendiculars) long by 21 feet 0.03 inches (6.410 metres) beam, 5 feet 11 inches (1.803 metres) draught displacing 400 tons / 462 tons (406 tonnes normal / 469 tonnes deep). As with *HMS Viper*, the construction of the hull was subcontracted to R. & W. Hawthorn Leslie and Company. She was laid down at their Hebburn-on-Tyne shipyard on 10th April 1901 and launched on 11th February 1902. As with *HMS Viper*, two sets of compound steam turbines were installed, with the high-pressure turbines directly driving the outer shafts and the low-pressure turbines the inner shafts, and two propellers per shaft. The Admiralty signed a contract for *Python* in May 1902, renaming the ship *HMS Velox* and she commissioned in February 1904. Originally Parsons were prepared to guarantee 32 knots but hoped for 33.75 knots, but *HMS Velox* was the first destroyer to be affected by the change of Admiralty requirements that speed trials should be carried out fully loaded rather than lightly loaded hence the guaranteed speed was reduced to 27 knots. During sea trials she achieved 34.5 knots light load and 27.142 knots average over the measured mile in full load. However, fuel consumption was significantly higher than expected. As-built two 150 shp (112 kW) triple expansion engines were installed and coupled to the inner shafts for efficient cruising, however cruising speed was only 10.3 knots and the contract requirement was 12 knots which was subsequently increased to 15 knots by the Admiralty – cruising turbines were substituted in 1907.

4.5 OTHER DESTROYERS (1904 TO 1910)

After *HMS Velox*, all Royal Navy Destroyers were turbine powered, such ships constructed during the period in question were:

- *HMS Eden* (part of 36 strong *E* or *River* Class two funnel Destroyers completed from 1904, rest of Class steam reciprocating engines) – 226 feet 6 inches (69.037 metres) long by 23 feet 9 inches (7.239 metres) beam, 7 feet 9 inches (2.362 metres) draught displacing 541 tons / 615 tons (550 tonnes normal / 625 tonnes deep). She was built by R. & W. Hawthorn Leslie and Company and laid down at their Hebburn-on-Tyne shipyard on 12th June 1902, launched on 13th March 1903, and commissioned in June 1904. Propulsion was five Parsons direct drive steam turbines, one high-pressure turbine on the centre shaft and two low-pressure turbines on the outer shafts producing 7,000 shp (5,220 kW) and 25.5 knots / 26.2 knots on trials. The two other turbines were coupled to the outer shafts and intended for cruising, *HMS Eden* was the first ship to be fitted with cruising turbines – a complex arrangement.
- *Cricket* Class two funnel Coastal Destroyers – 36 ships completed between December 1906 and September 1910. They ranged from 175 feet to 180 feet (53.340 metres to 54.864 metres) long by 17 feet 6 inches to 18 feet (5.334 metres to 5.486 metres) beam, 6 feet (1.829 metres) draught displacing 225 tons to 255 tons / 400 tons (229 tonnes to 259 tonnes normal / 406 tonnes deep). Propulsion was three Parsons steam turbines producing 3,750 shp (2,796 kW) and 26 knots.
- *F* or *Tribal* Class three to six funnelled Destroyers - 12 ships completed from 1907 to 1910. They were from 275 feet (83.820 metres) long by 28 feet (8.534 metres) beam, 8 feet 6 inches (2.591 metres) draught displacing 860 tons to 940 tons (874 tonnes to 955 tonnes). Propulsion was two or three Parsons steam turbines producing 12,500 shp (9,321 kW) and 33 knots

4.6 *HMS AMETHYST* (1905)

The Royal Navy ordered the three funnelled 360 feet (109.728 metres) long by 40 feet (12.192 metres) beam, 14 feet 6 inches (4.420 metres) beam displacing 3,000 tons (3,048 tonnes) *Topaze* Class (third-class) Protected Cruiser *HMS Amethyst* from Armstrong Whitworth. She was laid down at their Elswick shipyard in January 1903, launched on 5th November 1903 and was completed in March 1905, when she was the largest warship in the world powered by steam turbines. As with *HMS Eden*, propulsion was provided by five Parsons direct drive steam turbines, one high-pressure turbine on the centre shaft and two low-pressure turbines on the outer shafts producing 10,000 shp (7,457 kW) giving a top speed of 23.6 knots. Again, the two other units were cruise turbines coupled to the outer shafts. Compared to her three reciprocating engined sisters she was 1.3 knots faster and 10% more efficient.

From 1904, all major British warships used steam turbine propulsion.

4.7 *HMS DREADNOUGHT* (1906)

The two funnelled 527 feet (160.630 metres) long by 82 feet 1 inch (25.019 metres) beam, 29 feet 7.5 inches (9.030 metres) beam displacing 18,120 tons / 20,730 tons (18,410 tonnes / 21,062 tonnes deep) *HMS Dreadnought* was the world's first all big gun Battleship. Her armament and speed made all previous British and other nations Battleships obsolete overnight and hence triggered an arms race with every nation wanting steam turbine driven dreadnoughts in their fleet. She was laid down at HM Dockyard Portsmouth on 2nd October 1905, launched on 10th February 1906 and was commissioned on 2nd December 1906, when she was the largest and most powerful warship in the world. She was 20% (3.6 knots) faster than previous Battleships, equivalent to having 70% more power. Propulsion was provided by eight Parsons direct drive steam turbines, two high-pressure ahead and two high-pressure astern turbines on the outer shafts, and two low-pressure ahead turbines and two cruising turbines on the inner shafts. Rated performance was full speed 21 knots on 23,000 shp (17,151 kW) at 320 rpm with steam supplied at 250 psgi, and cruising at 14 knots on 6,000 shp (4,474 kW), although 27,000 shp (20,134 kW) was measured on trials and the maximum speed is 21.6 knots.

4.8 NON-BRITISH PARSONS TURBINE WARSHIPS (1902 TO 1907)

Although not strictly applicable in this section it is interesting to note the adoption of steam turbine propulsion, namely using Parsons machinery, in other countries during the period leading up to the completion of *RMS Mauretania*:

- 1902 – French two funnelled Torpedo Boat No. 293, built by Parsons Marine Steam Turbine Company. Three shaft / turbines as *Turbinia* directly driving individual shafts but only one propeller per shaft, produced 1,900 shp (1,417 kW) giving a top speed of 26.6 knots.
- 1904 – German two funnelled Torpedo Boats *S125* and *G137* (part of the *S90* Class) built by Parsons Marine Steam Turbine Company. As with *HMS Eden*, propulsion was five Parsons direct drive steam turbines, one high-pressure turbine on the centre shaft and two low-pressure (and two cruise) turbines on the outer shafts producing 6,600 shp (4,922 kW) and 28 knots.
- 1905 – German three funnelled 3,661 tonne Light Cruiser *SMS Lubeck* (part of the *Bremen* Class) built by Aktien-Gesellschaft Vulcan Stettin (AG Vulcan Stettin). Similar turbine arrangement to *HMS Amythest*, but four shafts not three, slower speed but longer range, 11,500 shp (8,576 kW) giving 22.5 knots.
- 1906 – French four funnelled Destroyer *Chasseur*. Three shafts, one high-pressure turbine directly driving the centre shaft and two low-pressure turbines directly driving the outer shafts, with a cruising turbine on the starboard shaft. Power was 9,000 shp (6,711 kW) giving a top speed of 30.6 knots.
- 1907 - From this year, all major French warships used steam turbine propulsion.
- 1907 – German three funnelled 3,814 tonne Light Cruiser *SMS Stettin* (part of the *Konigsberg* Class) built by AG Vulcan Stettin. Similar arrangement to *SMS Lubeck*, but 30% more power, 13,500 shp (10,067 kW) giving 25 knots.

5. STEAM TURBINE PROPELLED MERCHANT SHIPS (1900 TO 1908)

The previous Section discussed in some detail the development in steam turbine propulsion in warships. However, due to space limitation it is not possible to go in to as much detail for merchant ships, specifically passenger ships, over the same period, but the following give a brief outline of the developments in a general chronological order:

- The first steam turbine powered merchant ship was the two funnelled passenger ferry *TS King Edward* built by William Denny and Brothers in Dumbarton on the Clyde, and completed in 1901. She was 250 feet 6 inches (76.352 metres) long by 30.1 feet (9.174 metres) beam, 7 feet (2.133 metres) draught, with a tonnage of 502 GRT carried 1,966 passengers. She had three Parsons steam turbines, one high-pressure and two low-pressure, directly driving three shafts, one screw on the centre shaft and two on the outer ones (reduced to one in 1905), producing 3,500 shp (2,610 kW) giving 20.57 knots on trials.
- The two funnelled passenger ferry *TS Queen Alexandra* was built by William Denny and Brothers, and completed in 1902. She was 270 feet (82.296 metres) long by 32.1 feet (9.784 metres) beam, with a tonnage of 665 GRT. She had three Parsons steam turbines, one high-pressure and two low-pressure, directly driving three shafts, one screw on the centre shaft and two on the outer ones (reduced to one in 1905), producing 4,400 shp (3,281 kW).
- *TY Tarantula* was built by Yarrow and Company in Poplar on the Thames, and completed in 1902. She was based on the lines of contemporary first-class torpedo boats. She had three Parsons steam turbines directly driving three shafts, three screws on each like *Turbinia*.
- The two funnelled passenger ferry *TSS The Queen* was built by William Denny and Brothers, and completed in 1903. She was 309 feet 9 inches (94.411 metres) long by 40 feet

(12.192 metres) beam, with a tonnage of 1,676 GRT carried 1,250 passengers. She had two Parsons steam turbines directly driving three shafts giving 21 knots.

- The two funnelled passenger ferry *SS Brighton* was built by William Denny and Brothers, and completed in 1903. She was 273.6 feet (83.393 metres) long by 34 feet 2 inches (10.414 metres) beam, with a tonnage of 1,384 GRT. She had three Parsons steam turbines directly driving three shafts, one screw on each, giving 21 knots.
- The single funnelled yacht *SY Emerald* was built by Alexander Stephen and Sons in Govan on the Clyde, completed in 1903, and was the first steam turbine ship to sail across the Atlantic. She displaced 797 tons (810 tonnes), with a tonnage of 694 GRT. She had three Parsons steam turbines directly driving three shafts, one screw on each.
- The two funnelled passenger ferry *TS Princess Maud* was built by William Denny and Brothers, and completed in 1904. She was 300.6 feet (91.623 metres) long by 40.1 feet (12.222 metres) beam, with a tonnage of 1,746 GRT. She had three Parsons steam turbines directly driving three shafts giving 18.5 knots.
- The two funnelled passenger ferry *TS Londonderry* was built by William Denny and Brothers, and completed in 1904. She was 330.6 feet (100.767 metres) long by 42.1 feet (12.832 metres) beam, with a tonnage of 2,086 GRT carried 1,600 passengers. She had three Parsons steam turbines, directly driving three shafts, one screw on each, giving 22 knots on trials.
- The two funnelled passenger ferry *TSS Manxman* was built by Vickers and Sons and Maxim, Barrow-in-Furness, and completed in 1904. She was 341 feet (103.937 metres) long by 43 feet (13.106 metres) beam, 16 feet (4.877 metres) draught displacing 2,048 tons (2,081 tonnes), and carried 2,020 passengers. She had three Parsons steam turbines directly driving three shafts, producing 6,300 shp (4,698 kW) giving 22 knots.
- The single funnelled ocean liners *RMS Victorian* and *RMS Virginian* were built by Workman, Clark and Company in Belfast, and completed in 1905. They were 520 feet (158.496 metres) long by 60 feet (4,698 metres) beam, with a tonnage of 10,629 GRT carried 1,690 passengers. They had three Parsons steam turbines, one high-pressure and two low-pressure, directly driving three shafts, one screw on each, producing 12,000 shp (8,948 kW) giving a design of 17 knots but 19.5 knots on trials. These sisters were the first large turbine-powered commercial ships of any kind, and the first the three-shaft ships and turbine liners in Atlantic service.
- The two funnelled passenger ferry *SS Dieppe* was built by Fairfield Shipbuilding and Engineering Co. in Govan on the Clyde, and completed in 1905. She was 273 feet 5 inches (83.337 metres) long by 34 feet 8 inches (10.566 metres) beam, with a tonnage of 1,210 GRT. She had three Fairfield steam turbines directly driving three shafts, one screw on each, giving 22 knots.
- The two funnelled Cunard trans-Atlantic liner *RMS Carmania* was built by John Brown and Co. on Clydebank, and completed in 1905. She was 650.4 feet (198.242 metres) long by 72.2 feet (22.007 metres) beam, with a tonnage of 19,524 GRT carried 2,650 passengers. She had three Parsons steam turbines, one high-pressure and two low-pressure, directly driving three shafts, one screw on each, giving 18 knots. *Carmania* and her sister, the steam reciprocating propelled, *RMS Caronia* were the largest ships in the Cunard fleet prior to the introduction of *RMS Lusitania* and *RMS Mauretania*.
- The two funnelled passenger ferry *TS Dutchess of Argyll* was built by William Denny and Brothers, and completed in 1906. She was 250 feet (76.2 metres) long by 30.1 feet (9.174 metres) beam, 7 feet 4 inches (2.235 metres) draught, with a tonnage of 583 GRT. She had three Parsons steam turbines, one high-pressure and two low-pressure, directly driving three shafts, one screw on each, giving 21.65 knots on trials.
- The single funnelled passenger ferry *TS Atalantia* was built by John Brown and Co. on Clydebank, and completed in 1906. She was 210.4 feet (64.130 metres) long by 30.1 feet (9.174 metres) beam, with a tonnage of 486 GRT. She had three John Brown steam turbines directly driving three shafts, one screw on each.
- The two funnelled ocean liners *RMS Heliopolis* and *RMS Cairo* were built by Fairfield Shipbuilding and Engineering Co. in Govan on the Clyde, and completed in 1908. They were 525 feet 11 inches (160.299 metres) long by 60 feet 4 inches (18.390 metres) beam, with a tonnage of 11,117 GRT carried 1,114 passengers. They had three steam turbines, one high-pressure and two low-pressure, directly driving three shafts, one screw on each, producing 18,000 shp (13,422 kW) giving 19 knots.

6. ROYAL MAIL SHIP *MAURETANIA* (1907 TO 1935)

The Royal Mail Ship *Mauretania* was one of the most famous trans-Atlantic liners ever built. She was 787 feet (239.878 metres) long by 87 feet 6 inches (26.670 metres) beam, had a 33 feet 6 inch (10.211 metres) draught displacing 37,960 tons (38,571 tonnes), and with a tonnage of 31,938 GRT carried 2,165 passengers. Hence, she was the world's largest ship until the commissioning of White Star Line's *RMS Olympic* in 1911 as well as the fastest until the commissioning of Norddeutscher Lloyd's *SS Bremen* in 1929, which together with her sister *SS Europa* was also powered by Parsons steam turbines built under licence.

Mauretania had an older sister by two months, built by John Brown and Co. on Clydebank, namely *RMS Lusitania*. The only significant differences between these express trans-Atlantic liners was that *Mauretania* was three feet longer, 1 foot wider, had a slightly heavier lightship, had two extra stages of turbine blades in her forward turbines, and initially had cowl vents whereas *Lusitania* had 'oil drum' shaped vents which due to their fragility were gradually replaced.

These two sisters were conceived in 1902 when the Cunard Steamship Company decided to build two gigantic liners in order to win back the Blue Ribband from Germany, who had held it since Norddeutscher Lloyd's (NDL, North German Lloyd) *SS Kaiser Wilhelm der Grosse* had taken it from Cunard's *RMS Lucania* on 3rd April 1899 with an average speed of 22.29 knots.

The two funneled *RMS Lucania* had held the Blue Riband since 31st August 1894 with initially an average speed of 21.65 knots increasing it to 21.81 knots by 26th October 1894. She was built in 1893 by Fairfield Shipbuilding and Engineering Company on the River Clyde in Govan, Scotland, and was, with her sister *RMS Campania*, the joint largest passenger liner afloat when she entered service. She was 662 feet (201.778 metres) long by 65 feet 3 inches (19.965 metres) beam, with a tonnage of 12,950 GRT carried 2,000 passengers, and had two triple expansion steam reciprocating engines producing 31,000 shp (23,117 kW).

The *SS Kaiser Wilhelm der Grosse* was the first German four-funnel super-liner and was built by Aktien-Gesellschaft Vulcan Stettin (AG Vulcan Stettin) in 1897. She was 655 feet (199.644 metres) long by 65 feet 9.6 inches (20.056 metres) beam and displaced 23,300 tons (23,684 tonnes), with a tonnage of 14,394 GRT carried 1,506 passengers, and had two triple expansion steam reciprocating engines producing 33,000 shp (24,608 kW).

Hamburg-Amerikanische Packetfahrt-Actien-Gesellschaft's (HAPAG, Hamburg America Line) *SS Deutschland* took the Blue Riband on 12th July 1900 with an average speed of 22.42 knots, and had increased the record to 23.06 knots by 1st August 1901. She was built by AG Vulcan Stettin in 1900, was 679 feet 9 inches (207.188 metres) long by 67 feet 4 inches (20.523 metres) beam, 27 feet 11 inches (8.509 metres) draught displacing 26,918 tons (27,350 tonnes), with a tonnage of 16,703 GRT carried 2,050 passengers, and had two quadruple expansion steam reciprocating engines producing 37,000 shp (27,590 kW).

NDL's *SS Kronprinz Wilhelm* took the Blue Riband on 16th September 1902 with an average speed of 23.09 knots. She was built by AG Vulcan Stettin in 1901, 663 feet 3.6 inches (19.294 metres) long by 66 feet (20.117 metres) beam, 28 feet (8.534 metres) draught displacing 24,900 tons (25,299 tonnes), with a tonnage of 14,908 GRT carried 1,761 passengers, and had two quadruple expansion steam reciprocating engines producing 33,000 shp (24,608 kW).

Deutschland re-took the Blue Riband on 8th September 1903 with an average speed of 23.15 knots. *Deutschland* held it for the next four years until *RMS Lusitania* captured it for Cunard on 10th October 1907 with an average speed of 23.99 knots. She had considerably increased the record to 25.65 knots by 12th August 1909, when on 30th September 1909 *RMS Mauretania* took it with an average speed of 26.06 knots, and held it for the next two decades.

6.1 *RMS MAURETANIA*'S CONSTRUCTION (1904 TO 1935) AND SERVICE (1907 TO 1935)

Mauretania was designed by Cunard's naval architect Leonard Peskett. Initially she had a quadruple three bladed propeller installation, as shown in Figure 4, which was later changed to four bladed propellers, directly driven by Parsons steam turbines, two high-pressure on the outer shafts and two low-pressure (plus the astern turbines) on the inner shafts, developing a total of 76,000 shp (56,673 kW) at 180 rpm with steam supplied at 195 psi from 25 coal fired cylindrical multi-tubular boilers. This was, by some margin, the largest turbine installation at the time, and her machinery was constructed by the Wallsend Slipway and Engineering Company located just down river from her builders.

Ship 735, as she was called, was laid down on Number 1 Berth at Swan Hunter and Wigham Richardson's Wallsend shipyard on the north bank of the River Tyne on 18th August 1904, launched on 20th September 1906 by the Duchess of Roxburghe, see Figure 5, finally sailed from the yard on 22nd October 1907, as shown in Figure 20, and completed trials on 6th November 1907, as shown in Figure 27, achieving 25.73 knots, and was handed over and commissioned in to Cunard's fleet on 11th November 1907.

Figure 4: *RMS Mauretania* prior to launch from Number 1 Covered Berth at Swan Hunter and Wigham Richardson's Wallsend Shipyard [5]

Figure 5: Launch of *RMS Mauretania* from Swan Hunter and Wigham Richardson's Wallsend Shipyard on 20th September 1906 with Hebburn-on-Tyne and R. & W. Hawthorn Leslie's Shipyard in the background [5]

Mauretania departed Liverpool on her maiden voyage on 16th November 1907. She captured the record for the fastest eastbound crossing of the North Atlantic from her sister *Lusitania* on her return voyage in 5th December 1907, with an average speed of 23.69 knots, which *Mauretania* progressively increased to 25.88 knots by 21st June 1909. On 30th September 1909 she captured, with an average speed of 26.06 knots, the Blue Riband for the fastest, more difficult, westbound crossing, against the Gulf Stream, from *Lusitania* who has held it since October 1907. *Mauretania* went on to hold the Blue Riband for over two decades, a duration which was unsurpassed until 1970's – the current holder *SS United States* having held it since July 1952! *Mauretania*, and her older sister *RMS Lusitania*, were the only ships with direct-drive steam turbines, as fitted to the *Turbinia*, to hold the Blue Riband as later ships all utilised reduction-geared turbines.

Following service in World War One and after fire damage to the first-class cabins amidships and the first-class dining room floor above whilst alongside at Southampton on 25th September 1921, *Mauretania* returned to the Tyne shipyard of her birth in 1921 for a major overhaul, where her boilers were also converted to oil firing therefore increasing her developed power to 90,000 shp (67,113 kW) and hence allowing her to increase her average speed on the eastbound crossing to 26.25 knots on 25th August 1924.

Figure 6: *RMS Mauretania* leaving the River Tyne on 11th March 1922 following a six month overhaul and conversion to oil burning at Swan Hunter and Wigham Richardson [6]

Following losing the Blue Riband to Germany's *SS Bremen* on 22nd July 1929, her over 20 year old Parsons steam turbines were modified to produce more power, and as a result she increased her own speed records both eastbound and westbound however not by enough to match *Bremen* and *Europa*'s near 28 knot records.

Finally after 28 years of exemplary service, *Mauretania*'s final eastward crossing from New York to Southampton was in September 1934 and was completed at an average of 24 knots - the original contracted service speed specified three decades earlier, a true testament to her designers and builders. She was then laid up at Southampton alongside the former White Star Line flagship *RMS Olympic*. Her furnishings and fittings were auctioned off in May 1935 and on 1st July she departed Southampton en route for The Forth in Scotland to be broken up by Metal Industries at Rosyth [7 to 9]. However, on her way up the North Sea she stopped off the River Tyne where, as illustrated in Figure 7, thousands of people flocked to the seafronts of South Shields, Tynemouth and Whitley Bay to bid her farewell and rockets were fired from her bridge in a salute to her birthplace. After half an hour she resumed her northerly course, docking in Number 1 Drydock at Rosyth on 4th July, as shown in Figure 8. Demolition commenced on 21st August and was completed within twelve months.

Figure 7: *RMS Mauretania* stopped off the Tyne July 1935 on her way to the shipbreakers Metal Industries at Rosyth on the Firth of Forth [10]

Figure 8: *RMS Mauretania* moored afloat in Number 1 Dock at Rosyth awaiting the commencement of demolition [7]

Now the only physical representation of the *Mauretania* is in model form, an example of which is in Newcastle upon Tyne's Discovery Museum, as shown in Figure 9 below.

Figure 9: Builder's 1:48 model of *RMS Mauretania* as currently on display in the Tyneside Challenge gallery of Newcastle-upon-Tyne's Discovery Museum, with a 1:48 model of *SY Turbinia* in the right of the display case [Tyne and Wear Archives and Museums]

6.2 *RMS MAURETANIA*'S PRINCIPAL PARTICULARS (AS-BUILT 1907)

Mauretania's principal particulars are given in Table 1 below, and from these it should be noted that her dimensions, ratios and form are not uncommon for ships intended for an express liner service in the North Atlantic on a year-round basis hence combining good powering performance but also stability and seakeeping.

Table 1: *RMS Mauretania's* Principal Particulars

Particular	Imperial	Metric
Length Overall, L_{OA}	787' 0"	239.878 metres
Length Perps., L_{PP}	760' 0"	231.648 metres
Breadth mld., B_{MLD}	87' 6"	26.670 metres
Depth mld., D_{MLD}	60' 6"	18.440 metres
Draught mld., T_{MLD}	33' 6"	10.211 metres
Displacement, Δ	37,960 tons	38,571 tonnes
L_{PP}/B_{MLD} ratio	8.686	
Block Coeff., C_B	0.596	
Prismatic Coeff., C_P	0.615	
LCB (%L_{PP} aft of ⊗)	1.40%	

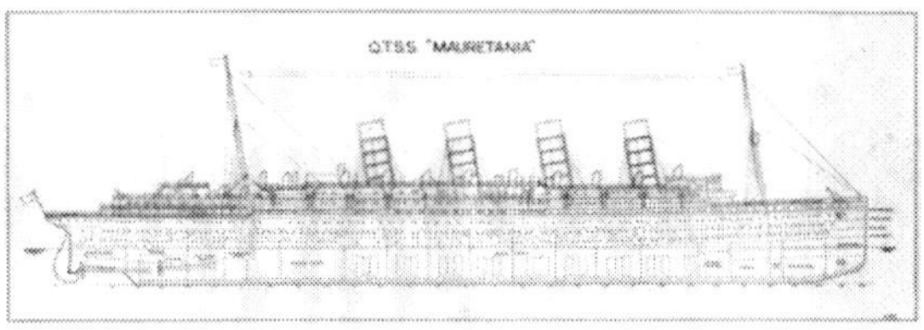

Figure 10: Profile of *RMS Mauretania*

6.3 *RMS MAURETANIA*'S HULLFORM (AS-BUILT 1907)

Work to refine the hull form was conducted in the Towing Test Tank at the Admiralty Experimental Works, AEW, at Haslar in Gosport, where *Turbinia* had been tested at the end of the previous century. It should be noted that the lines immediately forward of the rudder and the balanced rudder itself, which was of a special design and turned to 37 degrees, followed naval practice in order to improve the ships' turning response. Exhaustive experiments were also conducted over a period of two years in the non-tidal Northumberland Dock at East Howdon, two miles downriver from Swan Hunter and Wigham Richardson's Wallsend Shipyard, with a 47 feet 6 inch long model (launch), which was both towed and self-propelled utilising electric motors.

Figure 11: *RMS Mauretania* in Canada Dock, Liverpool, 1909 [11]

6.4 *RMS MAURETANIA*'S MACHINERY (AS-BUILT 1907)

Leonard Peskett's original 1902 design for Cunard's new trans-Atlantic liners, to become the *Lusitania* and *Mauretania*, incorporated reciprocating steam engines, three funnels and three shafts. However, Sir Charles Parsons proposed that the liners should be powered by his steam turbines rather than reciprocating engines. The advantages of steam turbines were:

- more powerful, reliable and efficient at 70% to 100% ratings, hence lower running costs;
- smaller, hence allowing for more passengers / cargo;
- less manpower;
- low noise and vibration;
- no oil in the steam;
- extraction of power from steam down to vacuum 1 psia (0.07 bara) rather than approximately 10 psia (0.7 bara).

However, early steam turbines did have some disadvantages:

- turbines and propellers ran at non-optimal speeds, hence 10% to 20% loss in efficiency;
- losses occurred at part load (cruising speeds for warships, not typically applicable to merchant ships) due to throttling losses and as turbine speed fell in proportion with ship speed;
- low rotating speed resulted in numerous stages and typically 1 million blades;
- astern turbines produced only approximately 70% of the ahead power.

As the advantages far outweighed the disadvantages for a trans-Atlantic liner in 1904 Cunard decided to change the original specification of steam reciprocating prime movers to Sir Charles Parsons' new steam turbines and hence a fourth funnel was added, in order to accommodate exhaust from additional boilers required to drive the steam turbines. A fourth propeller shaft was also added, as it was felt that the produced power could not be absorbed by three propellers. Other final modifications to the ship design prior to commencement of construction and keel laying in 1904.

As with the *Turbinia*, the propeller shafts were directly-driven without the use of reduction gearboxes, but with only one propeller per shaft. The outboard shafts were driven by the high pressure turbines with the low pressure exhaust steam from these driving the inboard turbines. Astern turbines were also fitted to the inboard shafts only.

Figure 12: *RMS Mauretania's* steam turbines, high pressure set in the foreground, in the Wallsend Slipway and Engineering Company's workshops [12]

Mauretania's inboard propellers were 12 feet 6 inches forward of the Aft Perpendicular, AP, whereas the outboard ones were 91 feet 9 inches forward. To improve efficiency the two outboard propellers rotated outwards while the two inboard propellers rotated inwards. As-built, *Mauretania* was fitted with three bladed propellers on all shafts, as shown in Figure 4, with a Diameter, D_P, of 17 feet, Pitch, P, of 15 feet 9 inches, and Blade Area Ratio, *BAR*, of 0.92.

During trials and service some vibration was noted in *Mauretania* at high speeds but not enough to concern the passengers. However, in *Lusitania* there was heavy vibration due to interaction of the propeller wakes which made the 2nd Class cabins uninhabitable and as a consequence strengthening members were added aft. Redesigned four bladed propellers were substituted to both ships soon after build to minimise vibration and increase efficiency. During *Mauretania*'s 28 year career her propellers were changed several times, see Figure 11, and as discussed previously her boilers and turbines were also modified and improved.

7. VIRTUAL *RMS MAURETANIA* (2016)

The simulation of the *Mauretania* dates back about ten years when it was initially discussed if a mathematical and visual model should be constructed to mark the centenary of her completion in 1907. However, due to the level of work it was only last year that this was undertaken [13]. The mathematical and visual model was constructed, calibrated and tested using information from various publications within the public domain [5, 7 and 14] and also original documents from her builders, Swan Hunter and Wigham Richardson, including particulars, plans, trial reports etc. [15 to 17].

A wireframe of the 3D model of the *Mauretania* is shown in Figure 18, part rendered 3D models in Figures 14 and 16 and a fully rendered ones in Figures 12, 15, 17 and 19. It should be noted that whilst the mathematical model is complete and fully tested, the 3D model is still under development and more detail is scheduled to be added.

Figure 14: Port forward view of part rendered 3D model of the 2016 virtual *RMS Mauretania*

Figure 13: Starboard upper view of the 2016 virtual *RMS Mauretania*

Figure 15: Port forward view of fully rendered 3D model of the 2016 virtual *RMS Mauretania*

Figure 16: Starboard forward view of part rendered 3D model of the 2016 virtual *RMS Mauretania*

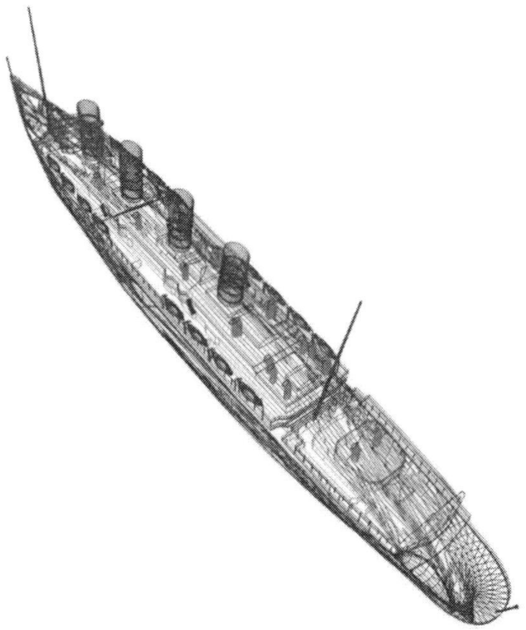

Figure 18: Port aft upper view of un-rendered 3D model of the 2016 virtual *RMS Mauretania*

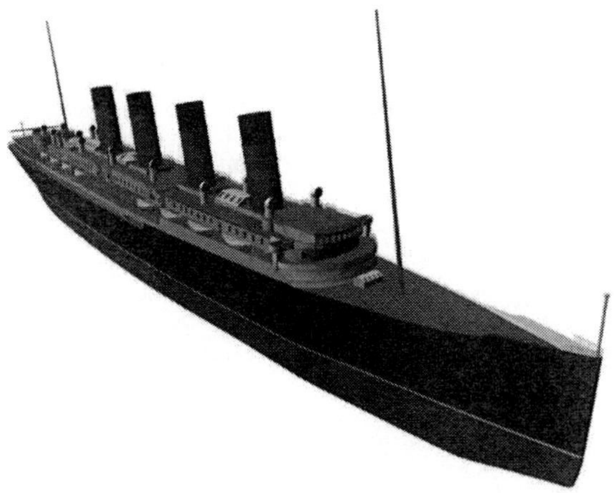

Figure 17: Starboard forward view of fully rendered 3D model of the 2016 virtual *RMS Mauretania*

Figure 19: Port aft upper view of fully rendered 3D model of the 2016 virtual *RMS Mauretania*

8. SIMULATIONS OF *RMS MAURETANIA* (2016)

By accurately modelling the hydrostatic and hydrodynamic properties of *Mauretania* in the Ship Simulator, the capabilities of which are described in [13] and to a lesser extent in [1], can be used to accurately predict and simulate how the ship would have responded to rudder, turbines and environmental factors. The *Mauretania*, and earlier *SY Turbinia* [1], models were novel projects that explored the limits of the simulator mathematical software.

The result of the simulation was that for the first time since *Mauretania*'s departure from the River Tyne in 1907, it was possible to recreate and experience how they would have handled and appeared to other ships as they sailed in the River Tyne and the North Sea, as illustrated in this section, or in fact any other port held as a geographical area within the South Tyneside College's Simulator system such as Liverpool, Southampton or New York [13]. Following are a few simulations of events that occurred over a century ago in September and October 1907 as if they happened today, December 2016.

8.1 *RMS MAURETAINIA* PRIOR TO DEPARTURE FROM HER BUILDERS

Figure 20 is the famous photograph taken on the 22nd October 1907 of the decade old and diminutive 31 metres long 43 tonnes *SY Turbinia* alongside her direct descendant the 240 metre long leviathan *RMS Mauretania*, the largest ship in the world. This shows the *Mauretania* at her builder's Outfitting Quay following her unofficial builder's sea trials in the North Sea the previous month and about to depart from the River Tyne for her official sea trials and handover to Cunard. Figure 21 shows the same foggy scene, as the Wallsend shoreline was in 1907, but this is not a real simulation as at this time *Turbinia* had mechanical problems and could not steam!

8.2 *SY TURBINIA* RIVER TRIALS

Figure 22 shows *Turbinia* [1] at speed heading towards the mouth of the River Tyne. In the background two passenger liners can be seen. These are *RMS Mauretania*, Swan Hunter and Wigham Richardson's Ship 735, and *MS Saga Ruby* (from 2014 the *MS Oasia*, and formerly between 1999 and 2004 *MS Caronia*), Swan Hunter Shipbuilders Ship 39 (built in 1973 as the *MS Vistafjord* for Norwegian America Line and the last true ocean-going passenger ship built in the United Kingdom). This is a fictitious scene, not only from the point of view of chronology, but because they are both moored at the former ship repairers Smith's Docks in North Shields, which to the best knowledge of the authors neither ship ever berthed.

8.3 *RMS MAURETAINIA* AND *SY TURBINIA* TYNE TRANSIT AND SEA TRIALS

Figure 23 is the famous photograph taken on 17th September 1907 of the *Mauretania*, leaving the River Tyne for her unofficial builder's sea trials. Note that, as discussed previously, *Turbinia* is not accompanying her as was planned due to her having mechanical problems. However, Figure 24 rectifies this and shows *Turbinia* escorting her descendant in to the harbour and Figures 25 and 26 show both ships leaving the Tyne and heading out in to the North Sea. Figures 27 and 28 are photographs of *Mauretania* off the Tyne on her unofficial builder's sea trials in September 1907 and then on her official high speed trial on 6th November 1907. Figure 29 shows both *Turbinia* and *Mauretania* simulated running at full speed north up the coast off the Tyne on the approach to the Whitley Bay measured mile, which was surprisingly used on 21st September 1907 at the end of *Mauretania*'s unofficial builder's measured mile trials [16 and 18] the rest having been conducted on the St. Abbs Head measured mile between the 18th and 20th September 1907.

Figure 20: *RMS Mauretania* and *SY Turbinia* at Swan Hunter and Wigham Richardson's Outfit Quay on the River Tyne at Wallsend 22nd October 1907 [12]

Figure 21: Virtual *RMS Mauretania* and *SY Turbinia* at her builder's outfitting quay, the former Swan Hunter and Wigham Richardson's as it was in 1907, on 'Long Reach' of the River Tyne at Wallsend December 2016

Figure 22: Virtual *SY Turbinia* at speed in 'Shields Harbour Reach' and 'The Narrows' of the River Tyne with the virtual *RMS Mauretania* and *MS Saga Ruby*, ex *MS Vistafjord* 1973, at the former Smith's Docks Quay at North Shields December 2016

Figure 23: *RMS Mauretania* leaving the River Tyne 17th September 1907 [12]

Figure 24: Virtual *RMS Mauretania* and *SY Turbinia* passing the Herd Groyne Lighthouse leaving 'Entrance Reach' of the River Tyne and entering the Harbour December 2016

Figure 25: Virtual *SY Turbinia* passing over the Bar and through the Piers at the Mouth of the Tyne at speed with *RMS Mauritania* in the Harbour December 2016

Figure 26: Virtual *RMS Mauretania* passing over the Bar and through the Piers at the Mouth of the Tyne at speed December 2016

Figure 27: *RMS Mauretania* under 'easy steam' off the River Tyne during mid-September 1907 [12]

Figure 28: *RMS Mauretania* on the Skelmorlie measured mile 6th November 1907 [5]

Figure 29: Virtual *RMS Mauretania* and *SY Turbinia* at speed off the River Tyne approaching the former Whitley Bay measured mile December 2016

9. CONCLUSIONS

This evolution of Ship Simulation systems has been driven by the desire to recreate an increasingly realistic training environment for operators of ship systems. Hand-in-hand with this complex requirement is the technological advances to make it possible, both in terms of hardware and software.

Undoubtedly it will be these factors, amongst others, that will see Simulators progress further into the future and hopefully, bring many historic ships 'back to life'. This would allow researchers etc. to explore first-hand and in real-time the performance of ships which are no longer seaworthy or in existence on full-scale commercial Ship Simulators. In addition, 'small scale' ship specific Simulators can facilitate the general public in experiencing operating and sailing on a preserved ship they are visiting and hence greatly enhance the experience and their knowledge.

Following on from the principal author's 2014 paper [1] to this forum, this paper has again demonstrated that the use of state of the art Simulation of *RMS Mauretania* as-built in 1907 is an invaluable tool in understanding the performance of historic ships.

10. DISCLAIMER

The views expressed in this paper are those of the authors and do not necessarily represent those of the organisations with which they are affiliated and the professional institutions of which they are members.

11. REFERENCES

1. HUTCHINSON, K.W. and HODGSON, P.V., '*SY Turbinia*– Genesis, Performance and Virtual Rebirth', *International Conference on Historic Ships 2014*, The Royal Institution of Naval Architects, London, United Kingdom, 25th to 26th November 2014. ISBN 978-1-909024-34-2

2. CLARKE, J., 'Charles Parsons – the man (1854 – 1931)', *Marine Propulsion - Turbinia and Beyond, International Conference*, The Institute of Marine Engineers, University of Newcastle upon Tyne, Newcastle upon Tyne, United Kingdom, 26th to 27th June 1997. ISBN 0-907206-87-5

3. SMITH, K., *Turbinia – The story of Charles Parsons and his Ocean Greyhound*, Tyne Bridge Publishing in association with Tyne and Wear Archives and Museums, Newcastle upon Tyne, United Kingdom, 2nd Edition, 1999. ISBN 978-1-85795-139-4

4. MACKIE, G.C. and HUTCHINSON, K.W., '*Turbinia* – state of the art? Then and now', *Marine Propulsion - Turbinia and Beyond, International Conference*, The Institute of Marine Engineers, University of Newcastle upon Tyne, Newcastle upon Tyne, United Kingdom, 26th to 27th June 1997. ISBN 0-907206-87-5

5. HOOD, A.G. and BOCLER, H, 'The Cunard Express Liner "*Mauretania*"', *The Shipbuilder*, London, United Kingdom, Volume II Special Number, November 1907.

6. SMITH, K., *Queens of the Tyne*, Tyne Bridge Publishing, Newcastle upon Tyne, United Kingdom, 2007. ISBN 978-1-85795-113-4

7. BUXTON, I.L., *Metal Industries - Shipbreaking at Rosyth and Charlestown*, World Ship Society, Kendal, United Kingdom, 1992. ISBN 0-915617-69-X

8. WILKINSON, M. and CORNWALL, T., *To the Breakers – the death of the "Mauretania"*, Amazon.co.uk, Ltd., Marston Gate, United Kingdom, 24th January 2014. ISBN 978-1-495311-85-7

9. KENT LAYTON, J. *The Unseen Mauretania 1907*, The History Press, Port Stroud, United Kingdom, 2015. ISBN 978-0-7509-5957-5

10. BLOWER, J., '*Mauretania* model set to spark bidding war at auction', *The Shields Gazette*, South Shields, United Kingdom, 30th April 2015.

11. Photograph 'John Currie – *Mauretania*', 1909. DS.WS/143/68, Tyne and Wear Archives and Museums, The Discovery Museum, Newcastle upon Tyne, United Kingdom

12. 'The Cunard Turbine-Driven Quadruple-Screw Atlantic Liner "*Mauretania*"' reprinted from *Engineering*, London, United Kingdom, 1907.

13. HUTCHINSON, K.W. and IRVING, M., 'The Simulation of Historic Ships – The record breaking Tyne built Victorian 'North Sea Greyhound' and her Edwardian Blue Riband 'Greyhound of the Atlantic' daughter', *13th International Conference on Marine Simulation*, The School of Marine Science and Technology, Newcastle University, Newcastle upon Tyne, United Kingdom, 8th to 11th September 2015. ISBN 978-1-909024-34-2

14. RICHARDSON, A., 'The Evolution of the Parsons Marine Turbine', *Engineering*, London, United Kingdom, 1911.

15. *Ship Particulars Book*, Swan Hunter and Wigham Richardson, Wallsend-on-Tyne, 1903 to 1911. DS.SWH/4/2/7/1, Tyne and Wear Archives and Museums, The Discovery Museum, Newcastle upon Tyne, United Kingdom.

16. *Sea Trials Book*, Swan Hunter and Wigham Richardson, Wallsend-on-Tyne, 26th January 1903 to 1st December 1909. DS.SWH/4/4/3/2, Tyne and Wear Archives and Museums, The Discovery Museum, Newcastle upon Tyne, United Kingdom.

17. *Specification for Turbine Machinery for Steel Quadruple-Screw Steamship*, The Cunard Steamship Company Limited, Liverpool, United Kingdom, 1905. DS/WS/97/1, Tyne and Wear Archives and Museums, The Discovery Museum, Newcastle upon Tyne, United Kingdom.

18. WHITEHEAD, I., 'Ships Speed Trials and the Measured Miles of the North East Coast', Joint Technical Meeting of The North East Branch of the Newcomen Society and The Discovery Museum, The Discovery Museum, Newcastle upon Tyne, United Kingdom, 4th December 2013.

12. AUTHORS BIOGRAPHIES

Keith W Hutchinson is Engineering Manager and Chief Engineer with Babcock Energy and Marine Technology at their Tyneside offices in North Shields, United Kingdom, where he undertakes commercial and offshore marine consultancy for a range of clients globally. He is responsible for the coordination and management of all engineering teams and, as senior Technical Authority, also for all governance aspects relating to whole Ship Design, engineering etc. He is Subject Matter Expert for the Naval Architecture domain across all offices in the United Kingdom.

Keith leads multidisciplinary teams formulating design solutions, together with cost estimating and investment appraisals. Over an almost 30 year career as a ship designer he has been constantly at the forefront of the technical development of robust but flexible concept / basic designs for a multitude of extremely novel / complex / bespoke merchant, offshore and naval vessels, both new-build and conversions - from sub 20 metre high speed catamarans / landing craft to FPSOs / FLNGs in excess of 500 metres for the harshest of environments - consistently striving with clients to explore unknown multiple dimensional design spaces.

Keith graduated in Naval Architecture and Shipbuilding from Sunderland Polytechnic in 1990 and from the University of Newcastle upon Tyne in 1992 with a Bachelor of Engineering in Marine Technology with First Class Honours in Naval Architecture.

Keith worked in a number of shipbuilding and shiprepair yards on the River Tyne, including Swan Hunter Shipbuilders, and upon joining his current employer in 1995, then Armstrong Technology, he was seconded to the Engineering Design Centre at Newcastle University for over three years as a Research Associate developing multiple criteria optimisation / decision support tools for the design of RO-RO passenger ships and other vessels.

Keith has published over 30 technical papers and articles on naval architecture and ship design etc., in addition to being invited to give many presentations. Keith sits on a variety of professional technical committees and is active in the organisation of a number of professional learned institutions. He is a Chartered Engineer, Chartered Marine Engineer, Fellow of the Royal Institution of Naval Architects, Fellow of the Institute Marine Engineering, Science and Technology and Fellow of the Society of Naval Architects and Marine Engineers.

Mel Irving is the Research and Development and Advanced Simulation Manager within the South Shields Marine School, part of South Tyneside College in South Shields, United Kingdom. Since 1861 the South Shields Marine School has been recognised as a world-renowned centre of excellence for marine education and training. The Marine Simulation Unit of South Tyneside College produces its' own area databases and ship models. He is responsible for the mathematical and hydrodynamic modelling of the ships used on the simulator system. He is involved in running all the project work which includes Port Development and feasibility studies, as well as the recreation of incidents for insurance purposes. His brief also includes the development of new business. He has been involved in various insurance claims, Port developments in Morocco, United Arab Emirates, Egypt, Orkney, Humber and Ullapool amongst many. Other work has included Civil Engineering projects such as the new Forth Road Bridge Crossing. He has also undertaken all simulation work for the Royal Navy's two new Queen Elizabeth Class Aircraft Carriers currently under construction.

Mel was educated at Carlisle Grammar School. In 1965 he joined Elder Dempster Lines of Liverpool as an indentured apprentice. This Company merged with Blue Funnel and Glen Lines becoming Ocean Fleets. During his apprenticeship he sailed on Passenger and General Cargo ships on routes to and from West Africa, the United States and the Far East. In 1969 he obtained his 2nd Mates certificate and joined A.P. Moller of Copenhagen. Here he served on board Crude Oil Tankers, General Cargo Ships and also on Product Carriers. He obtained his Masters Foreign Going certificate in 1976. In 1979 he left the sea and became a First Class Pilot in Orkney. The Pilotage work in Orkney was varied, from small coasters and sailing ships, warships, large Cruise Liners, to 350,000 deadweight tankers. For the last 15 years as a Pilot he was the Pilots representative, sitting on the examination body which was responsible for the Training and Licensing of authorised Pilots. He retired from piloting in 2008, whereupon he became involved with Pilot Training at South Tyneside College. This led to him becoming a full time member of staff.

Geoff Horseman is presently Chief Turbine Engineer at Siemens Power Generation Services at C.A. Parsons Works in Newcastle upon Tyne, United Kingdom. He is responsible for the technical direction of turbine work within the Engineering Department which includes the design of large steam turbines and auxiliaries for new power plant as well as upgrades for all makes of turbine and service support for the Siemens UK and Parsons fleets. Geoff also acts as the company historian and takes care of the Parsons archives. Geoff has been working at the Parsons site since 1980 and in steam turbine design since 1982.

AUTHORS' CONTACT DETAILS

Alexander de Vos
Traditional Boatyard "Klaas Hennepoel"
a.p.devos@hetnet.nl

Andrew Moyse
andrew.moyse1@ntlworld.com

Ben Mooiman
institution "Stichting Zr.Ms. Bonaire".
b.mooiman@wxs.nl

Chris Bailey
University of Greenwich
c.bailey@gre.ac.uk

Colin Bonnici
University of Malta
colin_bonnici89@hotmail.com

Diana McCormack
National Museum of the Royal Navy
diana.mccormack@nmrn.org.uk

Fred M Walker
fred.m.walker@btinternet.com

Hannah Cunliffe
National Historic Ships UK
Hannah.Cunliffe@Nationalhistoricships.org.uk

Ian Winkle
iewinkle@gmail.com

Jean-Baptiste R. G. Souppez
Southampton Solent University
jean-baptiste.souppez@solent.ac.uk

John Kearon
National Museums Liverpool
jkearon.maritime@hotmail.co.uk

Keith W Hutchinson
Babcock International Group
Keith.W.Hutchinson@babcockinternational.com

Luis Gabriel Alessio dos Santos
Southampton Solent University
luisg.alessio@gmail.com

Paul Handley
Paul Handley Design Ltd
paulhandley@onetel.com

Richard White
richard.white@mimer.no

Ron Neish
Maritime Training Academy, U.K.
theloftsman@gmail.com

Rosalind Spink
University of Hull
r.spink@solis-marine.com